Command and Control in U.S. Naval Competition with China

KIMBERLY JACKSON, ANDREW SCOBELL,
STEPHEN WEBBER, LOGAN MA

Prepared for the Office of the Secretary of Defense
Approved for public release; distribution unlimited

For more information on this publication, visit www.rand.org/t/RRA127-1

Library of Congress Cataloging-in-Publication Data is available for this publication.
ISBN: 978-1-9774-0536-4

Published by the RAND Corporation, Santa Monica, Calif.

Cover: U.S. Navy photo by Mass Communication Specialist 1st Class Shannon Renfroe

Preface

To help the Office of the Secretary of Defense better understand the balance of strategic competition between the United States and China, this report explores the following questions:

- How is command and control (C2) exercised in the U.S. Navy and China's People's Liberation Army Navy?
- How are these C2 concepts reflective of service culture?
- How do these C2 structures support or challenge each nation's shift to new maritime missions?

The report characterizes common themes that point to initial conclusions and identifies outstanding questions that merit greater exploration in both the topic of C2 and the ways that cultural and organizational factors influence China's defense choices more broadly.

This research was sponsored by the Office of the Secretary of Defense's Office of Net Assessment and conducted within the International Security and Defense Policy Center of the RAND National Defense Research Institute, a federally funded research and development center sponsored by the Office of the Secretary of Defense, the Joint Staff, the Unified Combatant Commands, the Navy, the Marine Corps, the defense agencies, and the defense intelligence enterprise.

For more information on the RAND International Security and Defense Policy Center, see www.rand.org/nsrd/isdp or contact the director (contact information is provided on the webpage).

Contents

Figure

Summary

As China pursues its rise as a global power, it is incrementally orienting its military toward power projection missions that support China's core territorial interests, as well as its expanding range of global missions. This gradual shift has particularly affected the People's Liberation Army (PLA) Navy, which is responsible for China's power projection efforts at sea. In response, the U.S. Navy must now also focus on enhancing its capabilities to conduct counter–power projection missions.[1] Although these mission sets are not entirely distinct—indeed, control of the seas enables power projection—this reality mandates the consideration of new strategies and capabilities. For both nations, these evolving postures require new doctrine and operational concepts, as well as myriad other changes affecting platforms, technology development, and personnel education and training.

Command and control (C2) in naval competition presents one lens with which to view these evolving missions. C2 is essential to the execution of power projection, as forces and platforms are often distributed and dispersed, and communications might be degraded or absent. Further, C2 models are reflective of organizational culture. The U.S. Navy and PLA Navy approaches to C2 are fundamentally different in part because of their dissimilar organizational cultures. This will have important implications as each country pursues its strategic goals through reoriented maritime missions.

---

[1] *Counter–power projection* primarily refers to sea control but is also inclusive of the U.S. Navy's distributed maritime operations concept, which envisions localized sea control operations supporting broader efforts at the fleet level.

Relying primarily on literature reviews, we explore the following questions: How is C2 exercised in the U.S. Navy and the PLA Navy? How do these C2 concepts reflect service culture? How do these C2 structures support or challenge each nation's shift to new maritime missions? Rather than answer these questions definitively, we intend for this report to be exploratory, identifying common themes and illuminating topics for future analysis.

Mission Command and the U.S. Navy

Mission command, a pillar of U.S. Navy culture for centuries, is central to its execution of power projection missions. This concept holds that leaders throughout the command chain are disciplined, fully apprised of their commander's intent, and empowered to make decisions and expediently execute actions to support their overall mission. Mission command requires transparency between all echelons and relies on subordinates understanding what to accomplish, rather than detailed direction about how to accomplish it.

In theory, mission command could be useful in counter–power projection efforts, as it enables an organization to develop technologically and conceptually, and ultimately innovate faster than its rivals. The U.S. Navy's recently developed distributed maritime operations concept is based on increasing the capability of individual platforms and spreading combat power out across long distances—increasing the ability of the total fleet to exercise sea control.

But mission command is challenging to execute in practice, given its reliance on open communication, trust, training, and discipline. Breakdowns in communication can occur, particularly under the stress of combat. Additionally, given that the U.S. Navy's primacy at sea has not been contested for decades, its ability to operate in a decentralized manner based on commander's intent has largely been untested.

U.S. Navy leadership believes that its C2 philosophy of mission command directly applies to its shift toward counter–power projection missions. However, success will likely require intensified cultivation of these principles through investments in education, trust, and professionalism across the force—in addition to reevaluating how it uses key platforms and related operating concepts and investment priorities.

The PLA Navy and Control and Command

Historically, the PLA Navy has utilized a C2 system that is fundamentally different from that of the U.S. Navy. Reflecting the Chinese Communist Party's authoritarian rule and overall PLA culture, the PLA Navy operates under tightly managed C2—better described as control and command—that allows for little delegation of authority or independent action. As the PLA Navy experiences more operations beyond China's coastal borders, this control and command structure has been increasingly stressed, causing problems, for example, with effective coordination between PLA Navy and Coast Guard vessels as they navigate evolving mission responsibilities.

In recent decades, China has demonstrated the ability to project significant maritime power in the Near Seas and an improved ability to project modest levels of naval power into the Far Seas.[2] However, these successes have been limited, noncombat missions with small flotillas. Exercising control and command over a larger number of vessels in a more distributed space in a combat environment offers far greater challenges.

The PLA Navy has also introduced a sizable number of new platforms, advances in information technology, and high-tech military weaponry, greatly enhancing its operational capabilities. Additionally, plans are in place to procure additional aircraft carriers. Moreover, evolving doctrine suggests the armed forces will be able to initiate and execute more-complex operations. However, these advances are occurring in a culture where superiors tend to micromanage their subordinates and tightly control tactical-level decisions, creating a much different C2 environment for power projection missions from that of the U.S. Navy. The PLA Navy's rigid control and command structure,

[2] The Near Seas are waters within the First Island Chain off of mainland East Asia, and the Far Seas are waters beyond the First Island Chain.

which endures even as its maritime operations have evolved, is likely to come under increasing strain given the relative independence and greater operations tempo required by power projection operations.

Topics for Future Study

The U.S. Navy and the PLA Navy are both likely to face challenges as they shift to new maritime missions unless they adapt their existing concepts of C2. Each navy's willingness to adapt could prove to be decisive in maritime competition, and perhaps ultimately in the overall balance of strategic competition between the United States and China. Currently, many unknowns exist, which point to several areas of inquiry for future research:

- **What is more valuable to China: the ability to project power globally or retaining its rigid control and command system?** These two goals are in inherent tension with one another, and one will likely have to give for the other to advance. However, we do not know where China is more likely to assume risk and what the ultimate implications on successful power projection will be.
- **Will the PLA Navy's increased experience and professional development affect the trust placed in PLA Navy personnel by senior PLA commanders? And how will increased PLA Navy professionalism affect control and command?** PLA Navy personnel are gaining additional operational experience at sea, theoretically building their professionalism and skills. However, will greater training and experience translate to greater trust in a subordinate's abilities to operate and think independently and execute the commander's intent?
- **Would the Chinese Communist Party tolerate a PLA Navy that is more empowered to make independent decisions?** A service that has greater autonomy and freedom of action is by definition more independent and powerful than one that is tightly and directly controlled. It is unclear whether China would tolerate such a greater concentration of power in the service and what

the implications would be for conducting maritime operations during wartime.

- **Would the PLA Navy taking a mission command approach to C2 be a threat to the United States?** An underlying assumption of this research was that a strict adherence to PLA's "control and command" approach would hinder the PLA's ability to achieve effective power projection. But would an adoption of mission command or an entirely different C2 system pose a greater threat?

Acknowledgments

First and foremost, we would first like to thank Andrew May of the Office of the Secretary of Defense's Office of Net Assessment for sponsoring this work and for continually supporting our research on topics that are as fascinating as they are challenging.

Further, we would like to thank Kristen Gunness and Cortez Cooper for providing careful reviews that improved the quality of this report. S. Rebecca Zimmerman, Derek Grossman, Jeffrey Engstrom, and Alice Shih also contributed to this study and helped shape the ultimate trajectory of our research. Additionally, we are grateful to Christine Wormuth, Michael Spirtas, and Michael McNerney for their oversight and guidance throughout the course of this project. Finally, we greatly appreciate Rosa Maria Torres for her administration support and efforts in preparing this report for final publication.

Abbreviations

C2	command and control
CCP	Chinese Communist Party
CMC	Central Military Commission
CNO	Chief of Naval Operations
DMO	distributed maritime operations
PLA	People's Liberation Army
PRC	People's Republic of China

CHAPTER ONE

Introduction

As China pursues its rise as a global power, it is incrementally orienting its military toward power projection missions that support both China's core territorial interests and its expanding range of global missions.[1] The Chinese Communist Party (CCP) has focused on growing and modernizing its People's Liberation Army (PLA) forces, advancing military technology, and expanding its ability to operate away from its borders through means such as the production of a second aircraft carrier and establishment of a PLA base in Djibouti. This gradual shift has particularly affected the PLA Navy, responsible for China's power projection efforts at sea. In response, the U.S. Navy, whose maritime dominance has remained largely uncontested since World War II, must now also focus on counter–power projection missions—specifically, sea control.[2] Although these mission sets are not entirely distinct—indeed, control of the seas enables power projection—this reality mandates the consideration of new strategies and capabilities. As the Chief of Naval Operations (CNO) stated in the 2018 *A Design for Maintaining Maritime Superiority*,

---

[1] While China's regional interests and territorial sovereignty remain its primary focus, power projection efforts away from Chinese shores are viewed as critical to protecting those nearer interests.

[2] Naval Surface Force, U.S. Pacific Fleet, *Surface Force Strategy: Return to Sea Control*, Naval Amphibious Base Coronado, Calif., 2017. *Counter–power projection* primarily refers to sea control but is also inclusive of the U.S. Navy's distributed maritime operations (DMO) concept, which envisions localized sea control operations supporting broader efforts at the fleet level.

> It has been decades since we last competed for sea control, sea lines of communication, access to world markets, and diplomatic partnerships. Much has changed since we last competed. We will adapt to this reality and respond with urgency.[3]

These are new strategic postures for both nations, each of which requires fundamental changes to their navies, not just in terms of platforms and technology but also in terms of new doctrine, operational concepts, institutional orientation, and personnel development. How the PLA Navy and the U.S. Navy shift to embrace these requirements will likely reflect their cultures and could have substantial implications for strategic competition between the United States and China.

Command and control (C2) in naval competition presents one particularly useful lens with which to view these challenges. C2 is an essential component of understanding how a nation executes power projection, as forces and platforms are often distributed and dispersed, and communications might be degraded or absent. In the U.S. Navy and in U.S. joint maritime operations more broadly,

> [t]he key tenets to command and control (C2) philosophy are the necessity of the subordinate commanders to execute operations independently but in accordance with a thorough understanding of the commander's intent, and command by negation or mission command.[4]

Mission command, or "the conduct of military operations through decentralized execution based upon mission-type orders,"[5] has been a pillar of U.S. Navy culture for centuries.[6] Mission command is

---

[3] U.S. Navy Chief of Naval Operations, *A Design for Maintaining Maritime Superiority*, Washington, D.C., Version 2.0, December 2018.

[4] Joint Publication 3-32, *Command and Control of Joint Maritime Operations*, Washington, D.C., Joint Chiefs of Staff, August 7, 2013, p. vii.

[5] Joint Publication 3-32, 2013, p. I-2.

[6] Colin Roberts, "The Navy," in S. Rebecca Zimmerman, Kimberly Jackson, Natasha Lander, Colin Roberts, Dan Madden, and Rebeca Orrie, *Movement and Maneuver: Culture and the Competition for Influence Among the U.S. Military Services*, Santa Monica, Calif.: RAND Corporation, RR-2270-OSD, 2019.

also central to the U.S. Navy's execution of power projection missions. This concept holds that leaders throughout the command chain are disciplined, fully apprised of their commander's intent, and empowered to make decisions and expediently execute actions to support their overall mission. The senior commander then has the ability to refute a subordinate's actions, retaining "command by negation." Mission command requires transparency between all echelons and relies on subordinates understanding what to accomplish, rather than detailed direction about how to accomplish it.

Joint Publication 3-32, which governs C2 in U.S. maritime operations, specifically states that commanders must "minimize detailed control" and expect that subordinates derive orders from their understanding of the purpose of the mission instead of "constant communications" with their commanders.[7] However, given the U.S. Navy's comparative lack of emphasis on sea control in the past several decades, it has been unclear to what extent the same philosophy is applicable to and executable in the service's counter–power projection efforts, particularly in today's competitive environment. Further, it is unclear whether the U.S. Navy's concept of mission command is necessarily the most effective C2 model for other nations, such as China, to use in their power projection efforts.

Historically, the PLA Navy has utilized a C2 system that is fundamentally different from that of the U.S. Navy. Reflecting the CCP's authoritarian rule and overall PLA culture, the PLA Navy operates under tightly managed C2—better described as control and command—that allows for little delegation of authority or independent action. In 2016, Chinese President Xi Jinping instituted far-reaching PLA restructuring intended to ensure the CCP's total control over the PLA. These reforms targeted, among other things, C2 processes and procedures in the PLA as its reach expands regionally and globally. How these reforms have affected the PLA Navy's C2 practices and, moreover, whether China can successfully execute power projection missions given its very different C2 philosophy remain largely unknown.

---

[7] Joint Publication 3-32, 2013, p. I-2–I-3.

Research Questions

In this report, we explore the following questions:

- How is C2 exercised in the U.S. Navy and the PLA Navy?
- How are these C2 concepts reflective of service culture?
- How do these C2 structures support or challenge each nation's shift to new maritime missions?

However, we do not attempt to definitively answer these questions, given the inherent difficulty in measuring cultural influence and the opacity of the Chinese military personnel system, among other factors. Rather, we characterize common themes that point to initial conclusions and identify outstanding questions that merit greater exploration in both the topic of C2 and in the ways that cultural and organizational factors influence China's defense choices more broadly.

This research nests within a broader topic of interest: What can we learn about Chinese priorities and potential future competitive advantages by understanding the strategic drivers that have influenced Chinese defense choices? A wide range of other potential research questions can be derived from this prompt, but we focus specifically on the insights provided by analysis of how C2 concepts are likely to be exercised in maritime competition.

Methodology, Limitations, and Organization of the Report

The research in this report was conducted primarily through literature reviews. Starting with a broad topic, our first step was to identify which aspect of military organization on which to focus. Using academic, U.S. government, and Chinese sources, we identified 15 Chinese defense behaviors, processes, and choices that contrast markedly with those of the United States.[8] These behaviors, processes, and choices

---

[8] The behaviors, processes, choices, and other aspects we initially identified were the role of the political work system, the role of Marxist theory, the evolution of the PLA's "active defense"

were compelling in that they have major bearing on Chinese military operations, because sufficient resources are available to enable analysis of such topics and that the topics remain relatively understudied. In consultation with the sponsor, we selected the topic of C2 concepts in power projection and counter–power projection. We then researched concepts of C2 in the PLA Navy and the U.S. Navy, in theory and in practice, and analyzed what effect they might have on maritime competition as both nations shift their respective naval missions.

This report is intended to be exploratory in nature and thus presents several limitations. First, although analysis of Chinese "hard power" components, such as weapon systems and platforms, is widely available, considerably less information exists on the less tangible aspects of the Chinese military, such as personnel management, education, and training. Additionally, many of the conclusions in this report are inherently speculative and will need to be tested, as we simply have not had substantial opportunity to observe the U.S. Navy's effectiveness in conducting sea control operations or the PLA Navy conducting robust power projection missions. Finally, as noted previously, our conclusions are designed to illuminate topics for future analysis, rather than to provide complete answers. Therefore, we believe that this report's primary purpose is to generate greater discussion about, and spur additional analysis into, the less tangible aspects of strategic competition between the United States and China.

In Chapter Two, we explore how mission command is practiced in the U.S. Navy, as well as its benefits and challenges, and consider how mission command might evolve as the U.S. Navy adjusts to counter–power projection missions. In Chapter Three, we analyze control and command in the PLA Navy, using four maritime operations cases to illustrate the challenges the service faces in its power projection efforts. Finally, in Chapter Four, we summarize our conclu-

national military strategy, the use of militia forces, the professionalization of the PLA, the role of noncommissioned officers in the PLA, the CCP's reliance on military operations other than war, the influence of Confucian thought on officer-enlisted relationships, PLA service culture, defense spending, the PLA's lack of overseas basing, the CCP's policies on human rights, the relationship between the citizens of the People's Republic of China (PRC) and the country's military, the PRC's role in peacekeeping, and the PLA's concept of C2.

sions and recommendations for areas of additional study to develop a greater understanding of how maritime competition between the United States and China might unfold in the future.

CHAPTER TWO

The U.S. Navy's Shift to Counter–Power Projection

For decades, the U.S. Navy has focused on power projection. This emphasis has become central to the U.S. Navy's institutional identity and is reflected in its organizations, strategies, and personnel processes. Given the U.S. Navy's historical and cultural emphasis on command at sea, its preferred C2 method, mission command, has become fully intertwined with its power projection missions.

As the U.S. Navy continues its efforts to better compete against adversaries, such as China, it is investing in ways to increase its sea control capabilities across all maritime domains. Although sea control and power projection are certainly not mutually exclusive, and indeed can be complementary, they elicit different sets of operational and strategic considerations. This might require a change in the way the U.S. Navy needs to approach C2, and such a change will almost certainly reflect U.S. Navy values and culture. As stated in U.S. joint doctrine, "C2 of maritime forces is shaped by the characteristics and complexity of the maritime domain, and the traditions and independent culture of maritime forces."[1]

In this chapter, we discuss how mission command reflects U.S. Navy culture, how it has been executed in the U.S. Navy's power projection missions over time, and the strengths and challenges of mission command for the U.S. Navy. We also explore the applicability of mission command to the U.S. Navy's shift to counter–power projection

[1] Joint Publication 3-32, 2013, p. I-2.

missions. How the United States approaches mission command in this counter–power projection mission will have implications for the PLA Navy as it shifts the focus of its naval forces more directly on power projection.

Culture and C2 in the U.S. Navy

In 1989, Carl H. Builder wrote that the U.S. Navy, "over anything else, is an institution" and that two primary aspects of Navy culture are its adherence to tradition and stature.[2] Its culture is also characterized by its defined hierarchies and divisions between and among officers and enlisted personnel and occupational specialties, with the unrestricted line communities presiding.[3] This emphasis on tradition is so strong that it has led some to conclude that the service might demonstrate "an unwillingness to adapt to changing circumstances."[4] The U.S. Navy is also defined by its valuation of independence, from the individual to the institution, which reflects its central emphasis on commanding at sea. These cultural attributes significantly shape the U.S. Navy's concept of C2 and have for centuries.

In the U.S. Navy's earliest years, to lead vessels that were separated by a great deal of physical distance from their higher echelon, a commander at sea needed a high degree of autonomy in decisionmaking because there were no modern forms of communication. As technology advanced and the character of warfare evolved, the centrality of command at sea to U.S. Navy culture did not change. If anything, technological change and the intensity of combined arms warfare increased the importance of sound, fast decisionmaking at lower echelons. At the outset of World War II, Admiral Ernest King wrote to his fleet:

[2] Carl H. Builder, *The Masks of War: American Military Styles in Strategy and Analysis*, Baltimore, Md.: Johns Hopkins University Press, 1989.

[3] Roberts, 2019.

[4] Roberts, 2019, p. 52.

> We are preparing for—and are now close to—those active operations (commonly called war) which require the exercise and the utilization of the full powers and capabilities of every officer in command status. There will be neither time nor opportunity to do more than prescribe the several tasks of the several subordinates (to say "what," perhaps "when" and "where," and usually, for their intelligent cooperation, "why"), leaving to them—expecting and requiring of them—the capacity to perform the assigned tasks (to do the "how").[5]

After World War II, however, the U.S. Navy was left searching for institutional relevance in a new era. The United States' primary adversary, the Soviet Union, posed little naval threat against which the U.S. Navy needed to prepare. Further, many believed that the invention of the atomic bomb meant the U.S. Navy's capabilities were antiquated, given that its ships were unable to defend against atomic weapons or launch an atomic bomb themselves.[6] To be sure, the U.S. Navy was not without a substantial role in strategic deterrence: Its nuclear-missile-armed submarine fleet did, and still does today, compose a major element of the nation's nuclear triad. Regardless, these realities turned the U.S. Navy's attention to advocating about the criticality of its aircraft carriers to power projection. As Builder observed in 1989, "The Navy could no longer claim to be the nation's first line of defense; but it could project significant power ashore by means of carrier-based aviation."[7]

Over the next few decades, the carrier became central to not only how the service organized and employed its forces but also its identity. Its primary focus was power projection, though the carrier also enabled sea control operations. However, the U.S. Navy's focus on power projection became almost singular following the end of the Cold War, when it became possible to presume sea control without contest based

---

[5] Ernest King, Commander in Chief, Atlantic Fleet, U.S. Navy, "Exercise of Command: Excess of Detail in Orders and Instructions," CINCLANT Serial 053, January 21, 1941.

[6] Builder, 1989.

[7] Builder, 1989, p. 78.

on the unparalleled warfighting prowess of the U.S. Navy.[8] Further, as Builder explained, "The capital ship, the aircraft carrier, and its supporting forces were justified by power projection; sea lane protection called for a different and less interesting Navy."[9] Today, this operational and cultural emphasis on the carrier endures. The U.S. Navy operates 11 nuclear-powered carriers capable of operating globally and supporting a wide variety of functions.

Due in part to the importance of the carrier to U.S. naval operations, the preference for operating forward was reinforced as a fundamental aspect of U.S. Navy culture. The U.S. Navy's historical C2 model of mission command, enabling forward-deployed commanders to make decisions derived from their commander's intent, proved conducive to this operating environment and thus became even more closely linked with conducting power projection missions. Today, the concepts of tradition, independence, operating forward, and mission command, all valued as central tenets of U.S. Navy leadership and culture, are foundational to how the U.S. Navy conducts power projection missions.[10] In 2019, Colin Roberts of the RAND Corporation wrote,

> In practice, independent action and initiative—whether independent command of a ship or squadron or simply independent initiative on the part of an action officer on a staff—form an important part of Navy ethos. . . . Waiting to be told what to do is anathema to Navy culture.[11]

---

[8] U.S. Navy Chief of Naval Operations, 2018, p. 3; Bryan Clark and Jesse Sloman, *Advancing Beyond the Beach: Amphibious Operations in an Era of Precision Weapons*, Washington, D.C.: Center for Strategic and Budgetary Assessments, 2016, pp. 6–8; Roberts, 2019, p. 55.

[9] Builder, 1989, p. 76.

[10] Roberts, 2019.

[11] Roberts, 2019, p. 53.

Mission Command in the U.S. Navy

Mission command relies on decentralized execution, which requires a high degree of trust within the chain of command, technical competency throughout all echelons, and empowerment of subordinates to make decisions. This assumes a high degree of professionalism, experience, and competence across the organization, and also that commanders are adept at articulating their intent. The relationship between the commander and the subordinate is of the utmost importance. The higher echelon trusts the lower-level commander to make certain decisions, and the commanders at sea exercise authority over the people in their charge and the technical systems at their disposal.

Orders emphasize what to do and why, but not *how*, at least beyond the minimum details necessary to ensure mission accomplishment.[12] One of the chief advantages of practicing mission command is that, in a warfare environment in which communications are degraded or denied, and/or decisions must be made rapidly, the ability to accelerate individual and collective decision cycles would theoretically enable a force to adapt more effectively to counter its adversary.[13] Of note, mission command does not necessarily preclude more detailed instruction where appropriate or certain authorities being centralized at higher echelons out of operational necessity.

The U.S. Navy's multiple warfare communities each share the same overall concept of C2, but, according to their different missions, organization, and equipment, the way in which they practice mission command can vary.[14] For example, the relationship between surface and subsurface platforms to their higher echelons requires commanders to generally operate with a high degree of autonomy, particularly in circumstances in which communications are degraded or nonexistent.

---

[12] Peter Vangjel, "Mission Command: A Practitioner's Guide," in Donald Vandergriff and Stephen Webber, eds., *Mission Command: The Who, What, When, Where, and Why: An Anthology*, Vol. 2, CreateSpace Independent Publishing Platform, 2019.

[13] John Boyd, "Organic Design for Command and Control," in Grant T. Hammond, ed., *A Discourse on Winning and Losing*, Maxwell, Ala.: Air University Press, 2018.

[14] Roberts, 2019, pp. 50–51.

By contrast, the centralized nature of joint air operations makes the relationship between tactical commanders and higher echelons in the aviation community far more stratified, with detailed command dominating out of necessity.

To help prepare its personnel to be able to execute operations effectively under this type of C2 structure, the U.S. Navy has highly developed systems for training, educating, and managing personnel. Education and training in proper procedures for operating at sea are inculcated at every level. Career progression is structured so that leaders grow in scope of responsibility and perspective as they progress through different positions.

Across the warfighting communities, the U.S. Navy concept of command leadership remains central to how the service develops leaders. The CNO promulgates a "charge of command" that communicates to the service what is expected of its commanding officers. This charge emphasizes the trust that must exist between the commander and crew, as well as superiors, to operate effectively.[15] These philosophical underpinnings serve as the foundation for the U.S. Navy's *Navy Leader Development Framework,* which shapes a strategic approach to developing Navy personnel into leaders capable of command at sea.[16]

This framework emphasizes warfighting competence, character, and personal connections. First, U.S. Navy leaders must be experts at their jobs and increase in skill as they grow. Second, behaving in accordance with the service's core values of honor, courage, and commitment engender the trust of the U.S. Navy's personnel. Finally, leadership is about connections, both personal and intellectual, that allow for the development of shared "mental models" and the ability to "anticipate [a] teammate's next move." The elements of mission command are apparent in the framework's logic and are integral to the U.S. Navy's concept of command.[17]

---

[15] John Richardson, *The Charge of Command*, Washington, D.C.: US Navy, April 6, 2018b, p. 1.

[16] John Richardson, *Navy Leader Development Framework,* Version 3.0, Washington, D.C.: U.S. Navy, May 2019.

[17] Richardson, 2019.

Strengths and Challenges of Mission Command in Counter–Power Projection Missions

In theory, mission command could be useful in counter–power projection efforts, as it enables an organization to develop technologically and conceptually and ultimately innovate faster than its rivals.[18] At the strategic level, constraining a competitor's ability to project power might depend on the ability to adapt more quickly to the new operating environment.[19] Indeed, service leaders have stressed the relationship between the tenets of mission command and the ability to adapt to a new environment, both in terms of organizational adaptability and the execution of the Navy's new operating approach.[20] Further, the philosophical approach of maneuver warfare holds that the force that can observe, orient, decide, and act the best and the fastest prevails in combat.[21]

Additionally, the recently developed DMO concept is based on increasing the capability of individual platforms and spreading combat power out across long distances. This approach, dubbed "distributed lethality," seeks to increase the ability of the total fleet to exercise sea control. The logic of this concept suggests that distributing the fleet—and, hence, delegating decisionmaking ability—does not necessarily detract from the ability to integrate its combat power.[22] This appears consistent with mission command.

U.S. Navy leaders have also stressed the need to create a more networked fleet to operate competitively in a counter–power projection

---

[18] Dale Moore and Gregory Smith, "The Navy Needs a Culture of Innovation," *Proceedings*, August 2019.

[19] Moore and Smith, 2019.

[20] U.S. Navy Chief of Naval Operations, 2018.

[21] Boyd, 2018.

[22] Lyla Englehorn, *Distributed Maritime Operations (DMO) Warfare Innovation Continuum (WIC) Workshop September 2017: After Action Report*, Monterey, Calif.: Naval Postgraduate School, Consortium for Robotics and Unmanned Systems Education and Research, November 2017; Naval Surface Force, U.S. Pacific Fleet, 2017, pp. 9–11.

environment.[23] This enhanced connectivity can enable information-sharing and the synchronizing of effects, both of which could theoretically be used for decentralized execution.

At the same time, however, a reliance on constant connectivity between echelons tends to support a directive command model and degrade the trust and empowerment needed for leaders to exercise initiative at lower echelons.[24] As a result, commanders and their staffs might tend to become more directive, relying less on mission-type orders that characterize mission command. Further, an overreliance on technical systems could erode warfighting competence by underemphasizing basic warfighting competencies and could further breed risk aversion.[25]

Additionally, some have argued that mission command might not be consistent with emerging operational concepts that rely on distribution and rapid execution,[26] such as multidomain operations, a U.S. Army concept with interest across the services and many common tenets with Marine Corps Expeditionary Advanced Base Operations, Air Force multidomain C2, and the Navy's DMO. Such concepts rely on synchronization to exploit "windows of advantage," both in terms of physical space and gaps in the adversary's decision cycle.[27] Leveraging technologically sophisticated systems and maneuvering in a synchronized manner requires a degree of timing that might not be possible in

---

[23] See, for example, John Richardson, "The Navy Our Nation Needs," speech delivered at the Heritage Foundation, Washington, D.C., February 1, 2018a.

[24] Graham Scarbro, "'Go Straight at 'Em!': Training and Operating with Mission Command," *Proceedings*, Vol. 145, No. 5, May 2019.

[25] Scarbro, 2019, pp. 2–3. See also Bing West, "American Naval Initiative—the Next Time Around," Hoover Institution, November 20, 2019. Retired U.S. Marine and author Bing West presents the opinion that distributed operations and limited ability to communicate demand a culture of decentralized execution.

[26] Andrew Hill and Heath Niemi, "The Trouble with Mission Command: Flexive Command and the Future of Command and Control," *Joint Force Quarterly*, Vol. 86, No. 3, June 21, 2017; Conrad Crane, "Mission Command and Multi-Domain Battle Don't Mix," *War on the Rocks*, August 23, 2017.

[27] U.S. Army Training and Doctrine Command, *Multi-Domain Battle: Evolution of Combined Arms for the 21st Century: 2025–2040*, Fort Eustis, Va., December 2017.

a culture grounded in the subordinate's initiative. Senior leaders across the joint force, however, have been explicit in the relationship between these concepts and mission command.[28] Such proponents of mission command contend that the risks of decentralization are mitigated by the shared understanding among dispersed actors that is fostered by this C2 construct and that the proper application of mission command enables rapid decisionmaking required to execute evolving concepts.

More generally, mission command can generate challenges irrespective of the missions in which it is exercised. First, successful execution of mission command is inherently difficult: It requires open communication, high levels of trust between commanders and their subordinates, and a high degree of training and discipline, all of which require significant time and investment.[29] Breakdowns in communication can occur, particularly under the stress of combat. When implemented irresponsibly, mission command can lead to disorganization. A lack of standardization of subordinate efforts can result in actions that are not aligned to a commander's intent. When the commander and the subordinate are not aligned, or the subordinate is not trained to such a standard to react and respond to changing circumstances appropriately, the risk of catastrophic errors increases.[30]

Further, the practice of mission command inherently deemphasizes the utility of centralized control and directive command, which could unnecessarily preclude their use in specific contexts where such systems might be useful.[31] Finally, because the U.S. Navy's primacy at

---

[28] Service leadership perspectives provided in Todd South, "The Army's Updated Warfighting Concept will Drive Its Formations, Planning, and Experimentation," *Army Times*, December 6, 2018; David Berger, *Commandant's Planning Guidance*, Washington, D.C.: U.S. Marine Corps, 2019; Amy McCullough, "Goldfein's Multi-Domain Vision," *Air Force Magazine*, August 29, 2018. The utility of mission command has been examined in, for example, Conrad Crane, "Mission Command and Multi-Domain Battle Don't Mix," *War on the Rocks*, August 23, 2017; Hill and Niemi, 2017.

[29] Donald E. Vandergriff, *Adopting Mission Command: Developing Leaders for a Superior Command Culture*, Annapolis, Md.: U.S. Naval Institute Press, 2019.

[30] L. Burton Brender, "The Problem of Mission Command," *Strategy Bridge*, September 1, 2016.

[31] Hill and Niemi, 2017; Crane, 2017.

sea was not contested for decades, the ability of its leaders to operate in a decentralized manner based on commander's intent has largely been untested, so there might be additional potential drawbacks to mission command in a power projection environment that we have not been able to observe.

Despite these drawbacks, proponents of mission command argue that a culture of decentralized execution is preferable to the alternative and that the drawbacks of mission command can be mitigated by conscientiously developing leaders to operate within such a culture.[32]

A Future Evolution of Mission Command in the U.S. Navy?

U.S. Navy leadership has made clear that the Navy's C2 philosophy of mission command directly applies to its shift toward counter–power projection missions. According to former CNO Admiral John Richardson,

> The current security environment demands that the Navy be prepared at all levels for decentralized operations, guided by commander's intent. This operating style is reliant on clear understanding up, down, and across the chain of command. It is also underpinned by trust and confidence created by demonstrating character and competence.[33]

However, how the service will need to evolve to support this revitalized mission remains an open question. In other words, what will the United States need to do to gain, or maintain, competitive advantage based on its exercise of mission command as it shifts to a new strategic posture?

To counter power projection, a force must rely on its "ability to deny partially or completely the enemy's use of the sea for military or

---

[32] Brender, 2016; Vandergriff, 2019.

[33] U.S. Navy Chief of Naval Operations, 2018, p. 7. The head of U.S. Navy surface forces has emphasized the same point, as well as the need to reinvest in sea control capabilities (Naval Surface Force, U.S. Pacific Fleet, 2017).

commercial purposes."[34] In its power projection missions, the U.S. Navy thought about sea denial in terms of countering an anti-access/area denial challenge: maintaining the ability to operate freely in certain areas despite technological systems and operating concepts designed to keep the Navy from doing so.[35] Under counter–power projection, this definition of sea denial expands beyond ensuring the U.S. Navy's ability to project power, to also focus on "the employment of forces to destroy enemy naval forces, suppress enemy sea commerce, protect vital sea lanes, and establish local military superiority in vital areas."[36]

In the absence of a salient challenge in recent decades, the U.S. Navy has not focused extensively on its sea control capabilities, as the service has been capable of striking nearly at will with standoff fires and aviation capabilities.[37] Although the U.S. Navy sees both sea control and power projection as interrelated components of an effective joint campaign,[38] as the counter–power projection mission grows in importance, the U.S. Navy's relative emphasis on sea control will likely have to increase. This shift, in turn, could demand a rethinking of the Navy's ability to practice mission command. Although the extent of the changes that the U.S. Navy will undergo to counter the power projection efforts of the PLA Navy remains to be seen, a shift to counter–power projection holds potential implications for four key aspects of how the service carries out mission command.

First, the U.S. Navy might need to reevaluate its reliance on the carrier strike group. The aircraft carrier has been a major investment priority for the U.S. Navy and at the core of its concept of fleet employment. The ability for embarked aircraft to exercise sea control while

---

[34] Joint Military Operations Department, *Syllabus and Study Guide for the Joint Maritime Operations Intermediate Lever Warfighter's Course*, Newport, R.I.: U.S. Naval War College, College of Naval Command and Staff and Naval Staff College, February 2018, p. 88.

[35] Sam Tangredi, "Antiaccess Warfare as Strategy," *Naval War College Review*, Vol. 71, No. 1, December 2018.

[36] Joint Publication 3-32, 2018, p. GL-7.

[37] Michael Bayer and Gary Roughead, *Strategic Readiness Review*, Washington, D.C.: U.S. Navy, December 3, 2017; Naval Surface Force, U.S. Pacific Fleet, 2017, pp. 1–2.

[38] Joint Publication 3-32, 2018, pp. I-3–I-4.

projecting power ashore made the aircraft carrier the logical centerpiece around which other capabilities could be organized. With control of the seas now contested, and the preferred methods of power projection now frustrated if not outright denied by advanced weapon technology, the way in which the carrier is used, and its value relative to other platforms, might change.[39] Long-range fires and advances in sensing capability have made the concentration of forces dangerous and ineffective.[40] Large-profile platforms are now challenged to operate close to an enemy's coastline, and U.S. Navy doctrine reflects the view that concentrating forces is no longer an effective way to exercise sea control.[41] The U.S. Navy will likely need to search for new ways to employ its carriers and, as its operating concepts indicate, change the size and composition of the force packages that it employs.[42]

Second, the relative importance of platform types and related operating concepts and investment priorities might shift. DMO calls for a fleet-centric approach to sea control and power projection.[43] Such a posture relies on enhancing the firepower of individual platforms and spreading them out, while integrating their combat capability. The lethality and survivability of each individual platform become more important,[44] as does the ability to network them (either through technical systems or, given a culture of mission command, through shared awareness and mission-type orders). Some have forecasted a shift to more small vessels rather than fewer, larger vessels.[45] Others have

---

[39] Bryan Clark, Adam Lemon, Peter Haynes, Kyle Libby, and Gillian Evans, *Regaining the High Ground at Sea: Transforming the U.S. Navy's Carrier Air Wing for Great Power Competition*, Washington, D.C.: Center for Strategic and Budgetary Assessments, 2018, pp. 1–4.

[40] Clark et al., 2018, pp. 1–4.

[41] Clark and Sloman, 2016, pp. 6–8; Englehorn, 2017, pp. 12–13.

[42] Naval Surface Force, U.S. Pacific Fleet, 2017, pp. 9–11.

[43] Englehorn, 2017, pp. 12–13.

[44] Naval Surface Force, U.S. Pacific Fleet, 2017, pp. 9–11.

[45] Ronald O'Rourke, *China Naval Modernization: Implications for U.S. Navy Capabilities—Background and Issues for Congress*, Washington, D.C.: Congressional Research Service, December 20, 2019, p. i.

emphasized how existing platforms will be leveraged in new ways (as in the previous example of the aircraft carrier).[46]

The U.S. Navy, as well as the broader national defense community, has highlighted the role of emerging technologies, such as unmanned systems, in providing additional sensing and combat power across long distances.[47] The extent to which the U.S. Navy as an organization will invest in and employ these new platforms and approaches is unknown, but shifts toward different warfighting capabilities, platforms, and even unmanned systems could have a profound impact on naval culture and its concept of C2 by fundamentally changing the way the service organizes itself and fights. However, this is not to say that all legacy approaches have outlived their utility. Aspects of the U.S. Navy's methods for sea control and power projection might still prove to be valid or be adapted rather than discarded.

Third, the U.S. Navy's view of conflict as a conventional military problem could be challenged. In the current operating environment, competitors, including the PRC, are challenging the United States asymmetrically with a range of instruments of national power below the threshold of conventional military conflict. The U.S. Navy has identified a "conceptual challenge" in dealing with this asymmetric competition.[48] If the U.S. Navy invests greater resources in such areas as interagency integration, irregular warfare, and building partner capacity, it would represent a departure from the Navy's traditional focus on securing resources for advancements in platform capacity and could also require different imperatives for effective C2.[49]

Finally, the U.S. Navy might have to refocus its training and education efforts to further support mission command and consider whether its current practices support the culture it needs. In 2017, the U.S. Navy noted systemic shortcomings in its leader development systems when it conducted a holistic assessment of U.S. Navy readi-

---

46 Clark et al., 2018, pp. 1–4.

47 Englehorn, 2017, p. 6.

48 U.S. Navy Chief of Naval Operations, 2018, p. 5.

49 Roberts, 2019, p. 57.

ness in the wake of the USS *Fitzgerald* and USS *John S. McCain* collisions. It observed that without the external pressure of a peer adversary, warfighting competence seems to have atrophied because of a lack of proper investment in training.[50] Proponents of mission command have noted that a decline in tactical competence, coupled with the expansion of staffs, tends to constrain the exercise of mission command.[51] According to this school of thought, decentralized execution is hindered by a decline in tactical competence, and a tendency toward more directive command. It is possible that the U.S. Navy might need to address the underlying structures, such as training, education, personnel management, and leadership, in general to reorient to the current competitive environment.[52]

Closely related to leader development are such issues as readiness, manpower, and maintenance. Since the late 2010s, the U.S. Navy has started to reevaluate its deployments and how it trains its personnel to undertake specific roles on ships.[53] This can influence how well prepared a crew is to execute its mission, particularly in the absence of directive orders. Additionally, certain concepts, such as the utilization of multiple, rotational crews on forward-deployed ships to address readiness challenges, could have bearing on the conduct of mission command. Such rotations would likely require even more widespread practice of mission command principles throughout the fleet, considering that a larger number of personnel would be practicing them in a given chain of command. Frequent rotations would also further emphasize the importance of trust between echelons and clarity of commander's intent, given that forces would have less time to establish relationships and develop "on-the-job" understanding in a new role.

Recent U.S. Navy policy moves have shown the service's interest in addressing these issues. For example, the service has stated that it

---

[50] Bayer and Roughead, 2017.

[51] Vandergriff, 2019.

[52] Bayer and Roughead, 2017.

[53] See, for example, David B. Larter and Mark D. Faram, "Mattis Eyes Major Overhaul of Navy Deployments," *Navy Times*, May 7, 2018; David B. Larter, "The U.S. Surface Fleet Is Shaking Up How It Readies Ships for Deployment," *Defense News*, May 6, 2019.

intends to leverage new technologies to increase training time and realism.[54] There has also been a high-profile effort at the secretarial level to overhaul the service's approach to education.[55] In his recent fragmentary order to Admiral Richardson, newly selected CNO Admiral Michael Gilday underscored the need for mission command, linking it to a need to incorporate "decision science" into U.S. Navy leader development.[56]

---

54 Darwin McDaniel, "Navy Looks to Improve Training with Live, Virtual, Constructive Technologies," *Executive Gov*, November 28, 2018.

55 Ben Werner, "Navy Proposing Bold Changes in How It Teaches Sailors in New Education Plan," *USNI News*, February 12, 2019.

56 U.S. Navy Chief of Naval Operations, 2018; Michael Gilday, *FRAGO 01/2019: A Design for Maintaining Maritime Superiority*, Washington, D.C.: U.S. Navy, December 2019, p. 6.

CHAPTER THREE

The PLA Navy's Shift to Power Projection

The CCP, the PLA, and the PLA Navy are defined by strict authoritarian rule, centralized control, and extremely limited delegation of authority. These factors shape organizational culture in the Chinese military and appear to preclude the practice of mission command, instead promoting a much more rigid concept of C2. As the PRC expands its power projection efforts, it will have to navigate the tension between its traditionally inflexible C2 concept and the demand for greater decentralization that is inherent to many power projection operations. Although mission command as practiced by the U.S. Navy is unlikely to be replicated exactly by the PRC, China will likely need to consider how to manage when platforms are geographically dispersed and communication between echelons is challenging. This is expected to have a substantial effect on the way the PLA Navy executes these missions, at least for the foreseeable future.[1]

Conducting this analysis is inherently challenging because we have little reliable firsthand information about PLA Navy command culture.[2] Additionally, much of the information we do have might not

---

[1] Roger Cliff, *China's Military Power: Assessing Current and Future Capabilities,* New York: Cambridge University Press, 2015. Cliff actually identifies "a fundamental mismatch between the PLA's doctrine and the organizational culture that would be required to effectively implement it."

[2] Indeed, culture continues to be an understudied dimension of the PLA. For some noteworthy exceptions, see Andrew Scobell, *China's Use of Military Force: Beyond the Great Wall and the Long March,* New York: Cambridge University Press, 2003; and Alison A. Kaufman and Peter W. Mackenzie, *Field Guide: The Culture of the Chinese People's Liberation Army,*

reflect ongoing changes, as the PLA Navy is currently undergoing an organizational transformation from a system that emphasizes individual services to one that stresses joint operations and integrated command.[3] However, drawing on noteworthy examples of China's maritime power projection from the past ten years, we can explore how the PLA's concept and practice of C2 appears to function in the contemporary PLA Navy and what challenges China could face as it continues its pursuit of global power projection.

In this chapter, we first analyze what is known about the organizational cultures and C2 processes of the PLA as a whole, as well as those of the PLA Navy. We then examine four contemporary examples of Chinese naval power projection missions to illustrate how the PRC's traditional concept of C2 might face challenges. Finally, we evaluate the impact of these missions on the PLA Navy's culture and assess the potential implications for the trajectory of the PLA Navy's future power projection efforts.

Culture in the PLA

The PLA, including the PLA Navy, possesses a culture that is fundamentally different from that of the U.S. military overall and that of the U.S. Navy specifically. Similarly, the PLA's command culture and the relationship between its leadership and China's top-level political and military leaders are also profoundly different from those in the United States. This difference is rooted in the distinctive traditions of political culture, civil-military culture, military culture, and service culture in the two countries.[4] The political culture in contemporary China is extremely hierarchical, and highly centralized organization—the

---

Alexandria, Va.: CNA, 2009. The latter report provides an extremely basic treatment of the PLA's organizational culture with no substantive treatment of the PLA Navy's culture.

[3] For the most comprehensive and up-to-date assessment of the transformation, see Phillip C. Saunders, Arthur S. Ding, Andrew Scobell, Andrew N. D. Yang, and Joel Wuthnow, eds., *Chairman Xi Remakes the PLA*, Washington, D.C.: National Defense University Press, 2019.

[4] The layered conception of culture is drawn from Scobell, 2003, pp. 2–7.

effect of an enduring traditional Chinese culture blended with a more recently imported Leninist culture and organization—was adopted by the CCP and is pervasive throughout the PRC.

The PRC's civil-military culture emphasizes complete obedience to and symbiosis with the CCP.[5] The PLA is first and foremost the CCP's army—and owes its loyalty to the party.[6] Military autonomy has tended to be either extremely weak or nonexistent, as the PLA is dependent on and intertwined with the CCP.[7] The CCP desires a PLA with a culture that is both "red" (politically loyal and ideologically pure) and "expert" (professionally competent and combat capable).[8] However, competence has suffered at times at the expense of loyalty to the party. Despite perennial CCP apprehension over the political reliability of the PLA, such concern seems misplaced.[9] Moreover, the PLA is certainly "expert," albeit not quite as proficient and operationally capable as the CCP and senior military leaders would like. Senior military leaders routinely refer to what have been called the "two incompatibles."[10] This means that the PLA (1) has yet to reach the level of modernization where it could be victorious in informatized war and (2) has yet to acquire the military capabilities to successfully undertake the operations it is supposed to execute.

The PLA has tended to have relatively weak individual service cultures—with the notable exception of the ground forces because of

---

[5] On the term *symbiosis,* used to describe civil-military relations in China, see Amos Perlmutter and William M. LeoGrande, "The Party in Uniform: Toward a Theory of Civil-Military Relations in Communist Political Systems," *American Political Science Review,* Vol. 76, No. 4, December 1982.

[6] On the primacy of the PLA's loyalty to the party and on the importance of loyalty and obedience in China's civil-military culture, see Scobell, 2003, pp. 57–58.

[7] See, for example, Sofia Ledberg, "Rethinking the Political Control and Autonomy of the PLA," paper presented at the CAPS-RAND-NDU PLA Conference, Taipei, Taiwan, November 15–16, 2019.

[8] Scobell, 2003, pp. 75–76.

[9] Andrew Scobell, *Civil-Military "Rules of the Game" on the Eve of China's 19th Party Congress,* Seattle, Wash.: National Bureau of Asia Research, October 2017.

[10] See, for example, Dennis J. Blasko, "The Chinese Military Speaks to Itself, Revealing Doubts," *War on the Rocks,* February 18, 2019.

a dominant "muddy boots" military tradition.[11] This is because services take on some of the cultural characteristics of the overall PLA. The PLA is steeped in a land power culture with its origins as a guerrilla force founded in the rural interior of China in the late 1920s—two decades prior to the formal establishment of the PRC in October 1949.[12] In contrast, the PLA's air and maritime forces were properly established only after the creation of the PRC—the PLA Air Force in November 1949 and the PLA Navy five months later.[13] For decades, the PRC Air Force and Navy have tended to operate as almost adjuncts to the PLA Army, each with relatively limited military capabilities, low political influence, and weak "service identities."[14]

Yet, as these services received greater funding and resources, especially since the 1990s, they began to develop a greater sense of service identity. This is especially true of the PLA Navy, which has taken a far more prominent role in China's national defense in the 21st century. These services assumed more-distinct individual identities as they earned official representation on the Central Military Commission (CMC) in 2004 and as their prominence was raised through the acquisition of high-profile platforms, such as the J-20 stealth fighter and aircraft carriers.[15] The relatively recent emergence of distinct ser-

---

[11] Michael S. Chase, Jeffrey Engstrom, Tai Ming Cheung, Kristen Gunness, Scott Warren Harold, Susan Puska, and Samuel K. Berkowitz, *China's Incomplete Military Transformation: Assessing the Weaknesses of the People's Liberation Army (PLA)*, Santa Monica, Calif.: RAND Corporation, RR-893-USCC, 2015.

[12] "The Communists' victory in 1949 was an Army victory, not a Navy one; the People's Liberation Army was unable to project power across even the narrow Taiwan Strait." Bernard D. Cole, *The Great Wall at Sea: China's Navy in the Twenty-First Century*, 2nd ed., Annapolis, Md.: Naval Institute Press, 2010, p. 7.

[13] The PLA Navy's headquarters was stood up in April 1950. Kenneth Allen, "Introduction to the People's Liberation Army's Administrative and Operations Structure," in James C. Mulvenon and Andrew N. D. Yang, eds., *The People's Liberation Army as Organization: Reference Volume v1.0*, Santa Monica, Calif.: RAND Corporation, CF-182-NSRD, 2002, pp. 21 and 23.

[14] Michael Chase, "The PLA and Far Sea Contingencies: Chinese Capabilities for Noncombatant Evacuation Operations," in Andrew Scobell, Arthur S. Ding, Phillip C. Saunders, and Scott W. Harold, eds., *The People's Liberation Army and Contingency Planning in China*, Washington, D.C.: National Defense University Press, 2015.

[15] See, for example, Chase, 2015, p. 65.

vice identities is potentially tempered by Xi Jinping's ambitious organizational initiative to emphasize jointness starting in 2016.

However, jointness does not simply become institutionalized in a matter of months or even years. Xi's organizational reforms, which remain a work in progress, have been dubbed "China's Goldwater-Nichols," and this is an apt analogy that holds relevant lessons for implementation in the Chinese context.[16] The landmark U.S. legislation passed in 1986, but it took many years to fully implement the act.[17] Entrenched U.S. service cultures were, and remain, resistant to jointness.[18] PLA service cultures are likely to be even more stubborn. For example, Chinese officers tend to serve their entire career in their home service, as joint appointments remain virtually unheard of in China.[19] This is reflective of the PLA's rigid professional military education system, although Xi's reforms are aiming to increase joint education and training in the PLA.[20] Moreover, stovepiping is a serious barrier to cooperation and coordination between bureaucracies, whether within the military or across civil-military boundaries. Stovepiping has

---

[16] Phillip C. Saunders and Joel Wuthnow, "China's Goldwater-Nichols? Assessing PLA Organizational Reforms," *Joint Force Quarterly*, Vol. 82, No. 3, July 2016.

[17] Public Law 99-433, Goldwater-Nichols Department of Defense Reorganization Act of 1986, October 1, 1986.

[18] S. Rebecca Zimmerman, Kimberly Jackson, Natasha Lander, Colin Roberts, Dan Madden, and Rebeca Orrie, *Movement and Maneuver: Culture and Competition for Influence Among the U.S. Military Services*, Santa Monica, Calif.: RAND Corporation, RR-2270-OSD, 2019.

[19] In a remarkable development, as of 2019 the land-locked Western Theater Command has a PLA Navy officer assigned to its staff. This reportedly is a very recent development. Phillip C. Saunders, "Beyond Borders: PLA Command and Control of Overseas Operations," paper presented at the CAPS-RAND-NDU PLA Conference, Taipei, Taiwan, November 15–16, 2019, p. 11, footnote 22.

[20] Joel Wuthnow and Phillip C. Saunders, *Chinese Military Reforms in the Age of Xi Jinping: Drivers, Challenges, and Implications*, Washington, D.C.: Center for the Study of Chinese Military Affairs, Institute for National Strategic Studies, National Defense University, March 2017.

been particularly problematic in the PRC: Chinese bureaucratic systems, or *xitong*, are notoriously insular, especially the military system.[21]

PLA Navy Culture

Historically, the PLA Navy has lacked a distinctive service identity and culture, having existed for decades in the shadow of the PLA's ground force. Therefore, its culture is likely synonymous with, or a close variant of, the PLA's overall organizational culture. Hence, similar to that of the PLA more generally, the PLA Navy likely exhibits a C2 structure that reflects the constrictive authoritarianism that characterizes the CCP's rule.[22]

Although the PLA Navy has more power projection experience than the other services, which could theoretically lead to a more progressive culture because of that experience, little evidence exists that its culture is notably distinct from that of the broader PLA, whose culture emphasizes control above command.[23] However, during the past two decades, the PLA's control and command culture has been stressed by the PLA Navy, given its expanding operations, and might have been forced by necessity to adapt and evolve to execute successfully a growing range of power projection missions in the Near and Far Seas. As shown in Figure 3.1, the Near Seas are waters within the First Island

---

[21] China has a long history of bureaucracies: the first professionally staffed government bureaucracy was established in the Han dynasty (206 BCE–220 CE), and today the PRC has one of the largest defense establishments in the world. One expert characterizes the PRC's military *xitong* as "virtually a state within the state." Kenneth Lieberthal, *Governing China: From Revolution Through Reform*, 2nd ed., New York: W. W. Norton, 2004, pp. 229–230. Writing in 2009 in the *Liberation Army Daily*, one PLA general officer complained: "The stove pipe issue is perhaps the greatest challenge we [i.e., the PLA] currently face." Cited in Andrew Scobell, "Discourse in 3-D: The PLA's Evolving Doctrine, Circa 2009," in Roy Kamphausen, David Lai, and Andrew Scobell, eds., *The PLA at Home and Abroad: Assessing the Operational Capabilities of China's Military*, Carlisle, Pa.: U.S. Army War College, 2010, p. 112.

[22] U.S. Office of Naval Intelligence, *The PLA Navy: New Capabilities and Missions for the 21st Century*, Suitland, Md., 2015; U.S. Defense Intelligence Agency, *China Military Power: Modernizing a Force to Fight and Win*, Washington, D.C., 2019.

[23] U.S. Defense Intelligence Agency, 2019.

Figure 3.1
Chinese Maritime Operations Areas

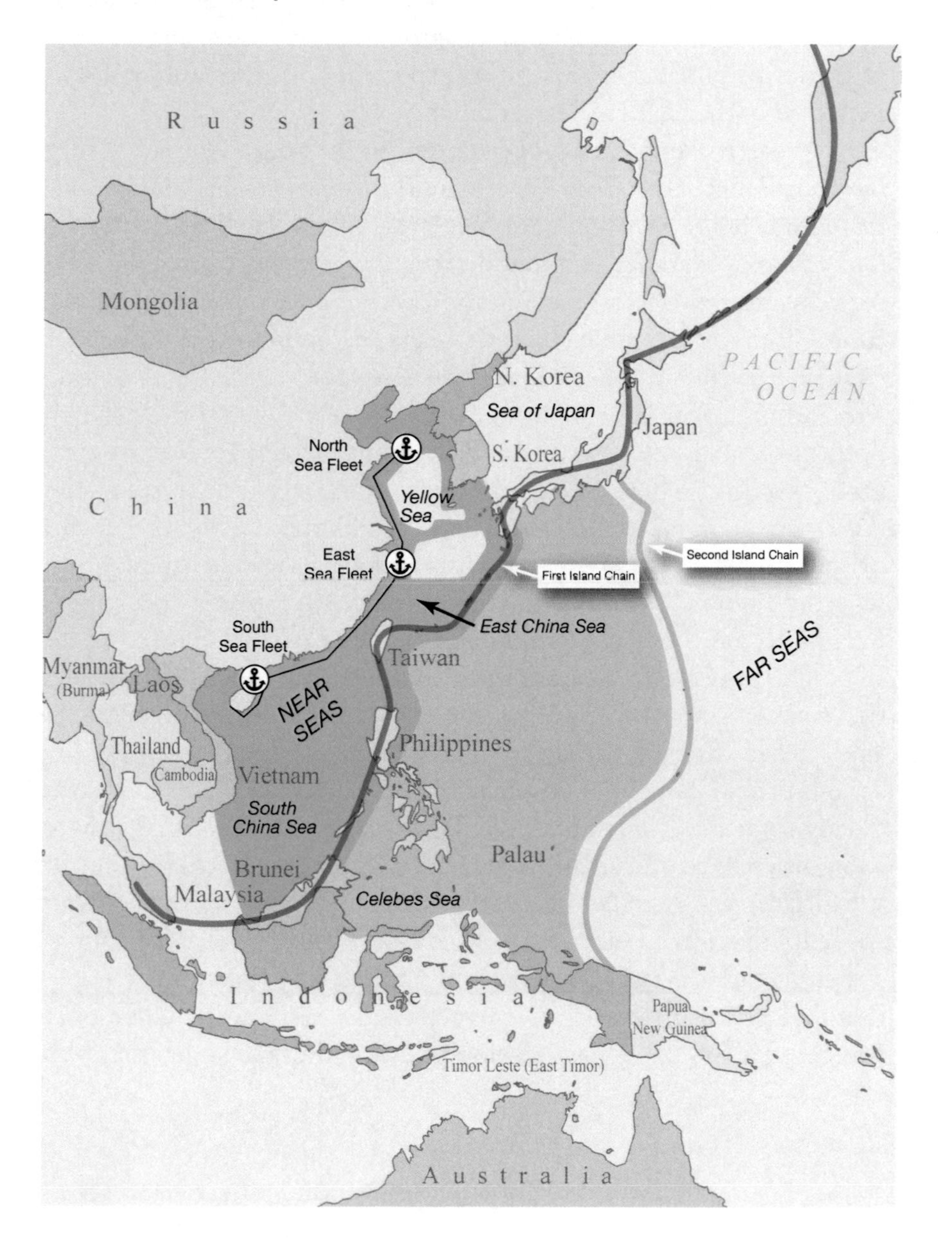

Chain, while Far Seas are waters beyond the First Island Chain.[24] The PLA Navy has been pushing limits beyond the other services not only because of its ambitious blue-water aspirations and growing inventory of platforms but also because of its expanding missions and operations tempo in both the Near Seas and Far Seas.

In particular, Near Seas operations in the South China Sea might have challenged the PLA Navy's traditional stovepiped bureaucratic insularity. Some potential examples that likely challenged the PLA Navy's rigid chain of command structure are the South Sea Fleet's battle to ensure effective coordination with the Coast Guard and maritime militia in defense of China's maritime interests, Far Seas operations in the Indian Ocean, and successive counter-piracy task forces. Meanwhile, Near Seas–plus operations (naval operations occurring both inside and just beyond the First Island Chain) by aircraft carrier groups could also have pushed the limits of the legacy C2 system, albeit less frequently and at a lower operational tempo. Lastly, the opening of the PLA's first official overseas naval base has broken new ground regarding how C2 should function at a PLA Navy's Far Seas port—meaning a naval base located outside the First Island Chain.

This process of adaption and evolution has reportedly spawned the emergence of a distinctive, nascent naval culture. In 2009, Fan Yihua, then-director of the PLA Navy's Political Department, penned a high-profile call to "promote the greater development and greater glory of navy culture" in the CCP's top theoretical journal.[25] Moreover, different naval subcultures could be emerging, including surface, subsurface, and marine, because of different missions and increased specializations. For example, the PLA Navy Marine Corps is tripling in size, from two to seven brigades, and dramatically expanding its operational reach to include greater expeditionary and amphibious operations, and it could begin to develop its own cultural characteristics that

[24] The First Island Chain refers to island formations stretching from the Kuril Islands to the Japanese archipelago, the Ryukyus, Taiwan, the Philippines, and Borneo.

[25] Fan Yihua [范印化], "Vigorously Promote the Greater Development and Greater Glory of Naval Culture [大力推进海军文化大发展大荣誉]," *Qiushi*, June 30, 2009.

are different from those of the broader PLA Navy.[26] While we believe this to be so, more research on the topic is necessary. Another question that merits further research is, how will these emerging subcultures affect C2 and China's overall ability to project power and compete with its adversaries?

The Concept of Control and Command in the PLA Navy

The PLA emphasizes *control above command*. Thus, it is appropriate to reverse the traditional order of *command and control* when examining the Chinese concept of C2 in the PRC and within the PLA.[27] Superiors keep subordinates on tight leashes, and individual initiative is not encouraged or rewarded.[28] Officers are not used to operating with a high degree of autonomy and tend to be uncomfortable making decisions independent of or outside their chain of command.[29] The PLA is beset by a culture of risk aversion characterized by low levels of trust in subordinates.[30]

Prior to Xi's organizational reforms, the military chain of command ran from the CMC through the PLA's General Staff Department and Military Region command staffs. Multiple additional C2 mechanisms were utilized to employ the PLA Navy's three fleets and

---

[26] See, for example, Xie Zhijun [谢志军] and Li Wei [李伟], "An Initial Examination of Establishing Culture with Marine Corps Special Characteristics [海军陆战队特色文化建设初探]," *Political Work Journal* [政工学刊], No. 11, 2017. This journal is published by the Dalian Naval Academy under the auspices of the PLA Navy's Political Department. See also Michael Peck, "China Is Tripling the Size of Its Marine Corps," *National Interest*, November 2, 2019.

[27] This point seems to hold for bureaucracies and military organizations in other Leninist systems. See, for example, Andrew Scobell and John M. Sanford, *North Korea's Military Threat: Pyongyang's Conventional Forces, Weapons of Mass Destruction, and Ballistic Missiles*, Carlisle, Pa.: U.S. Army War College, Strategic Studies Institute, 2007, pp. 10–13.

[28] John W. Lewis and Xue Litai, *Imagined Enemies: China Prepares for Uncertain War*, Stanford, Calif.: Stanford University Press, 2006, p. 146. See also Cliff, 2015.

[29] Cliff, 2015. See also Ledberg, 2019.

[30] According to one U.S. expert: "PLA culture is very risk averse." According to another U.S. expert: "[A] general fear of failure pervades [the PLA officers'] approach to work." Both are quoted in Cliff, 2015, p. 178.

their component vessels. The CCP's influence is built directly into the chain of command, including through political commissars and CCP committees, as well as the powerful General Political Department.[31] However, a fundamental goal of Xi's organizational reforms was to streamline PLA chains of command, including the establishment of the PRC's first joint staff entity—the Joint Staff Department.[32] Its predecessor, the General Staff Department, had been essentially a general staff for the PLA's ground forces.

The PLA Navy has remained ensnared in this wider control and command Chinese military culture: PRC naval vessels had tended to operate mostly offshore and in the Near Seas, where constant, real-time communication with onshore superiors was the norm. However, as PLA Navy vessels have ventured more frequently into waters farther from Chinese shores, this constricted culture of control and command has become increasingly strained.[33] This strain will manifest further as commanders at sea weigh the need to make swift judgment calls or delay decisions awaiting consultations with the PLA Navy staff back in Beijing. The first circumnavigation of the globe by a PLA Navy destroyer in 2002 was a landmark event, but the most striking and enduring control and command challenges faced by the PLA Navy in recent years have been posed by a range of complex missions.

Formally, the PLA Navy is charged "with the strategic requirements of near seas defense and far seas protection."[34] Currently, the

---

[31] On the General Political Department, see Larry M. Wortzel, "The General Political Department and the Evolution of the Political Commissar System," in James C. Mulvenon and Andrew N. D. Yang, eds., *The People's Liberation Army as Organization: Reference Volume v1.0*, Santa Monica, Calif.: RAND Corporation, CF-182-NSRD, 2002.

[32] Joel Wuthnow and Phillip C. Saunders, "Introduction: Chairman Xi Remakes the PLA," in Phillip C. Saunders, Arthur S. Ding, Andrew Scobell, Andrew N. D. Yang, and Joel Wuthnow, eds., *Chairman Xi Remakes the PLA*, Washington, D.C.: National Defense University Press, 2019.

[33] Guo Yuandan [郭媛丹], "Foreign Forces Engage in Close Reconnaissance of Chinese Ships During Far Seas Training, Express Pleasant Journey [中国海军远海训练遭外军抵近侦察被祝航行愉快]," *Global Times*, August 4, 2015.

[34] The most recent reiteration of this is contained in China's 2019 defense white paper. See State Council Information Office of the People's Republic of China, *China's National Defense in the New Era*, Beijing: Foreign Languages Press, July 2019, p. 28.

PLA Navy's primary power projection missions comprise two in the Near Seas and one in the Far Seas. The two main missions in the Near Seas are to (1) project naval power as part of a joint campaign across the Taiwan Strait in furtherance of its vital role in the PLA's sacred mission to unify China, and (2) coordinate with PRC paramilitary components, preferably below the threshold of war, to defend PRC territorial sovereignty and advance PRC maritime rights and interests in the East China Sea and the South China Sea.

The PLA Navy's main mission in the Far Seas is to protect the PRC's expanding overseas interests, most of which are well beyond the First and Second Island Chains. The main PLA Navy platforms for these missions are a wide range of surface and subsurface platforms, including destroyers, frigates, corvettes, and aircraft carriers, as well as submarines.[35] At the core of China's 21st century fleet are more than 30 destroyers, 50-plus frigates, and more than 40 corvettes.[36] In addition, the PLA Navy currently has two aircraft carriers in service, although as of December 2019 only one is operational (at least two or three more carriers are planned.) It is estimated that China currently possesses as many as 50 diesel-powered attack submarines. The PLA Navy also has approximately 60 amphibious ships, and more than 80 small patrol boats. These platform totals do not include China's fleet of approximately 250 capable Coast Guard vessels.[37]

---

[35] See, for example, Ronald O'Rourke, "PLAN Force Structure: Submarines, Ships, and Aircraft," in Phillip C. Saunders, Christopher Yung, Michael Swaine, and Andrew Nien-Dzu Yang, eds., *The Chinese Navy: Expanding Capabilities, Evolving Roles*, Washington, D.C.: National Defense University Press, 2011.

[36] The statistics in this paragraph are drawn from O'Rourke, 2019.

[37] O'Rourke, 2019.

Control and Command in the PLA Navy's Power Projection Missions

To understand how the PRC's traditional concept of control and command in the PLA Navy is exercised in practice, we examine four example peacetime missions.

Example A: Near Seas

Not surprisingly, the PLA Navy is the military service with the most experience in recent decades projecting power beyond China's shores. The most sustained and ongoing challenge has come in operations closest to home: in the Near Seas. The most challenging dimension of control and command in waters such as the East China and South China Seas has been interagency coordination with the Coast Guard and maritime militia. According to a June 2014 *Liberation Army Daily* article, the PLA is "actively building a military-police-civilian joint defense mechanism" and "closely coordinating with maritime forces to implement joint rights protection" in the Near Seas.[38] This is clearly a significant hurdle for the PLA Navy, and knowledgeable Chinese "stress that much more training is need for that scale of operation [described below] to be seamlessly executed on a regular basis."[39]

As a result of the PLA's recent organizational reforms, the PLA Navy operating in the Near Seas has a more streamlined chain of command, which runs from the CMC to the Joint Staff Department to the theater command to the fleet.[40] The Northern Fleet's chain of command flows through the Northern Theater Command with responsibility for the East China Sea, while the Southern Fleet's chain of command flows through the Southern Theater Command with responsibility for

---

[38] Cited in Linda Jakobson, "The PLA and Maritime Security Actors," in Phillip C. Saunders and Andrew Scobell, eds., *PLA Influence on China's National Security Policymaking*, Stanford, Calif.: Stanford University Press, 2015, p. 315.

[39] Jakobson, 2015, p. 316.

[40] Ian Burns McCaslin and Andrew S. Erickson, "The Impact of Xi-Era Reforms on the Chinese Navy," in Phillip C. Saunders, Arthur S. Ding, Andrew Scobell, Andrew N. D. Yang, and Joel Wuthnow, eds., *Chairman Xi Remakes the PLA*, Washington, D.C.: National Defense University Press, 2019, p. 137.

the South China Sea. At least in theory, interagency coordination has been simplified since Xi's recent organizational reforms. Under the restructuring effort the Coast Guard has been folded into the People's Armed Police, which now reports solely to the CMC.[41] Meanwhile, maritime militia forces are reportedly "managed by local . . . military commands," meaning that control and command functions appear to be handled by PLA Navy entities.[42]

Chinese military analysts have observed that this change will facilitate smoother execution of what has been named the "cabbage strategy" in the South China Sea.[43] This strategy involves extensive coordination between the PLA Navy, the China Coast Guard, and the maritime militia in which the militia in fishing boats and other ostensibly civilian vessels function as frontline forces backed by a second cabbage layer of Coast Guard vessels, with a third layer of PLA Navy vessels in reserve. Sometimes China's gray hulls operate over the horizon, and other times these gray hulls are in plain sight—albeit in a backup role.[44] The latter PLA Navy role was evident in May 2014, when seven gray hulls formed an inner ring of protection around a PRC oil rig, and dozens of maritime paramilitary vessels ("blue hulls") formed outer rings of protection some 120 miles off the coast of Vietnam.[45] This "highly coordinated maneuver was well executed" and presented an overwhelming display of force, and a confrontation with

---

[41] This was official effective January 1, 2018. See Joel Wuthnow, *China's Other Army: The People's Armed Police in an Era of Reform*, Washington, D.C.: Center for the Study of Chinese Military Affairs, Institute for National Strategic Studies, National Defense University, April 2019, p. 11.

[42] Andrew S. Erickson and Conor M. Kennedy, "China's Maritime Militia: What It Is and How to Deal with It," *Foreign Affairs*, June 23, 2016. The precise chain of command is unclear because the maritime militia is also managed by People's Armed Forces Departments—these are PLA organs embedded with local government bureaucracies. Dennis J. Blasko, *The Chinese Army Today: Tradition and Transformation in the 21st Century*, 2nd ed., New York: Routledge, 2012, pp. 40–42.

[43] Wuthnow, 2019, p. 23.

[44] McCaslin and Erickson, 2019, pp. 147–148. Gray hulls are naval ships.

[45] Carl Thayer, "China's Oil Rig Gambit: South China Sea Game-Changer?" *The Diplomat*, May 12, 2014.

some 30 Vietnamese ships passed without major incident.[46] This strategy has also been evident as part of the PLA South Sea Fleet Maritime Garrison District's "Three Warfares" training, when PLA Navy and China Coast Guard forces rehearsed how to operate together to deter foreign ships from Chinese waters.[47]

Available evidence suggests that the PLA Navy is capable of its "Near Seas defense" mission. That said, the primary control and command challenge confronting China's navy inside the First Island Chain is continued interagency coordination with the Coast Guard and the maritime militia. In this effort the PLA Navy has been "leading from behind," or letting the Coast Guard be the frontline force in the Near Seas.[48] Nevertheless, ensuring further successful applications of the cabbage strategy will likely require sustained effort and continuous training. Such operations have a far greater chance of being successful in the South China Sea than they do in the East China Sea. This is because while the coast guards of Southeast Asian states tend to be far less well trained and equipped than their PRC counterparts, the coast guards of Japan and South Korea are far more capable and outfitted.[49] If China's Coast Guard and maritime militia can continue to demonstrate the ability to execute Near Seas missions, it helps free up the PLA Navy to focus more on Far Seas power projection.

Example B: Near Seas "Plus"

One of the most dramatic examples in recent years of China's trajectory toward an ever-expanding reach in naval power has been the emergence of a full-throttle aircraft carrier program. This program is one of

---

[46] Andrew Scobell, "The South China Sea and U.S.-China Rivalry," *Political Science Quarterly*, Vol. 133, No. 2, 2018. The oil rig was eventually moved away from this location.

[47] Li Yimin and Li Xuefeng, "South Sea Fleet Aviation Division Relies on Information System to Boost the Level of Three Warfares Training," *Renmin Haijun*, September 2, 2014, p. 3c.

[48] McCaslin and Erickson, 2019, p. 147.

[49] Lyle J. Morris, "Blunt Defenders of Sovereignty: The Rise of Coast Guards in East and Southeast Asia," *Naval War College Review*, Vol. 70, No. 2, Spring 2017, pp. 88, 94, and 103. Morris focuses in particular on the capabilities of Japan's coast guard.

the clearest indicators of the seriousness of the PRC's intent to develop a significant Far Seas PLA Navy. The highly publicized formal commissioning of Beijing's first aircraft carrier, the *Liaoning*, in September 2012 signaled the dawn of a new era in PRC naval development.[50]

The fact that the *Liaoning* officially remains a training vessel and that when judged from a U.S. Navy perspective is not yet fully operational should not obscure the reality that the PLA Navy is gaining valuable experience in carrier operations and more Chinese-built carriers are in the pipeline to be commissioned in coming years.[51] While the *Liaoning* has primarily operated within the Near Seas, it has ventured out beyond the First Island Chain twice in recent years. The first time was in late 2016, when it circumnavigated Taiwan after exiting the Yellow Sea.[52] The *Liaoning* was escorted by two frigates and three destroyers—at least one vessel having been detailed from each of the PLA Navy's three fleets. The flotilla entered the Western Pacific via the Miyako Strait south of Okinawa and north of Taiwan, engaging in maneuvers east of the island, before heading back across the First Island Chain through the Bashi Strait some 90 miles south of Taiwan. Then, two and a half years later, the *Liaoning* again passed through the Bashi Strait. In this June 2019 passage, the carrier was once again accompanied by a flotilla of gray hulls.

These operations demonstrate the PLA Navy's intention to project power beyond the First Island Chain on a more routine basis. Nevertheless, the length of time between the two excursions suggests that progress toward this goal will be drawn out. Moreover, as of December 2019, China officially has only two operational aircraft carriers, as the newest, the *Shandong*, was officially commissioned on December 17,

---

[50] Andrew Scobell, Michael McMahon, and Cortez A. Cooper III, "China's Aircraft Carrier Program: Drivers, Developments, Implications," *Naval War College Review*, Vol. 69, No. 4, Autumn 2015.

[51] The *Liaoning* does not operate with the same capabilities and operational tempo of a U.S. Navy carrier: Flight operations are limited, and the vessel does not function as a fully integrated platform in the PLA Navy. Appropriately, the PLA Navy still classifies the *Liaoning* as a training vessel rather than a full-fledged warship.

[52] Ankit Panda, "Power Plays Across the First Island Chain: China's Lone Carrier Group Has a Busy December," *The Diplomat*, December 27, 2016.

2019.[53] The *Shandong* will reportedly be based at the PLA Navy's base at Sanya on Hainan Island and be part of the Southern Fleet.[54] Moreover, routine carrier presence to the Second Island Chain is at least five to ten years away. In addition to the *Liaoning*, another aircraft carrier is undergoing sea trials, at least one other is currently under construction, and as many as three more carriers are on the way.[55]

The control and command challenge these aircraft carriers present is an intra–PLA Navy one since the theater commands are not in the chain of command—the chain of command from the CMC's Joint Staff Department to the Navy Staff to the carrier group is direct.[56] Significantly, the most senior uniformed officer recorded as being present at the commissioning of the *Shandong* was General Li Zuocheng, who is the chief of the Joint Staff Department.[57] The main issue is one of coordination with other vessels to form a cohesive carrier group, including a special class of destroyers. According to Beijing-based naval expert Li Jie,

> It's quite an urgent need for the Chinese navy to have carrier group commanders—like its Western counterparts do—who are capable of commanding different warships and aircraft in modern joint-operation combat situations.[58]

The goal is to be able to operate the carrier group in the Far Seas beyond the range of the protective umbrella provided by land-based aircraft and missile defense.[59]

---

53 Minnie Chan, "First Made-in-China Aircraft Carrier the Shandong Officially Enters Service," *South China Morning Post*, December 17, 2019b.

54 Chan, 2019b.

55 Mark Episkopos, "Carrier Superpower? How Many Aircraft Carriers Does China Really Want?" *National Interest*, August 15, 2019.

56 Chan, 2019b.

57 Chan, 2019b.

58 Minnie Chan, "Chinese Navy Trains Top Guns to Command Expanding Aircraft Carrier Fleet," *South China Morning Post*, December 11, 2019a.

59 McCaslin and Erickson, 2019, p. 132.

Efforts to increase the range of China's aircraft carriers and exercise control and command structures for operations of naval carrier groups beyond the First Island Chain bear close monitoring because these are perhaps the single best way to gauge the PLA Navy's progress on blue-water power projection. To properly execute its "Far Seas protection" mission, the PLA Navy must enhance its ability to project a substantial package of platforms—well beyond what the PLA Navy to date has dispatched into the Indian Ocean (see "Example C: Far Seas"). It is too soon to assess the PLA Navy's progress and how its control and command culture is affecting operational performance, but monitoring the frequency and scope of these carrier group exercises beyond the First Island Chain could provide key insights. One such example could be the April 2018 carrier battle group exercise in the Western Pacific, which was executed in part to strengthen "the ability of commanders to make decisions when faced with complicated circumstances."[60] Although the extent to which these exercises are effective in training PLA Navy forces to operate far from Chinese shores is unknown and merits future research, an increasing number of exercises focused on these concepts could suggest greater familiarity across the PLA Navy officer corps with principles of C2 in a distributed posture.

Example C: Far Seas

The most significant and sustained new experience for the PLA Navy in recent years has been its counter-piracy operations in the Gulf of Aden. The first naval vessels for the mission were dispatched from China in December 2008, and a three-ship task force has been operating continuously ever since, with new vessels and fresh crews being rotated approximately every six months. This mission has been challenging to the PLA Navy in a multitude of ways. Some of these challenges are logistical in nature, such as refueling and providing food and fresh water. Another challenge has been maintaining the morale of Chinese sailors who are not accustomed to being at sea for many weeks

[60] "Carrier Leads Exercise in West Pacific," *China Daily*, April 23, 2018.

without port visits or shore leave.[61] However, while logistics remain a key challenge to the PLA Navy's efforts in the Far Seas, the service has made strides in addressing some of these issues. According to Andrew S. Erickson and Austin M. Strange, deployments in the Gulf of Aden enable strategists to study logistics needs of the PLA Navy and drove efforts to expand China's fleet of replenishment oilers.[62] Meanwhile, China has expanded port access in the region, including the 2017 establishment of China's first overseas base in Djibouti.[63]

These out-of-area operations have also proved challenging for the PLA Navy's control and command system.[64] On one hand, the chain of command has been streamlined—running from the PLA Navy staff directly to the task force, avoiding any theater command.[65] On the other hand, it is important to note that streamlining does not necessarily mean less top-level direction, only fewer steps between most senior and most junior levels. Further, such a high-profile Far Seas mission could make the CMC and its Joint Staff Department more prone to micromanage the task force. Since power projection missions in the Far Seas are still considered the exception rather than the rule, senior leaders could tend to meddle heavily in operational decisions rather than leave them to the task force commanders because they are seen to have critical strategic implications. In addition, there have been multiple coordination challenges, whether it is dealing with the navies of other countries patrolling the waters or communicating with the commercial

---

[61] Andrew S. Erickson and Austin M. Strange, *No Substitute for Experience: Chinese Anti-Piracy Operations in the Gulf of Aden*, Newport, R.I.: Naval War College China Maritime Studies Institute, November 2013.

[62] Andrew S. Erickson and Austin M. Strange, *Six Years at Sea . . . and Counting: Gulf of Aden Anti-Piracy and China's Maritime Commons Presence*, Washington, D.C.: Jamestown Foundation, June 2015, pp. 27–28, 97.

[63] Bill Chappell, "China Reaches Deal to Build Military Outpost in Djibouti," *Two-Way* (NPR), January 21, 2016.

[64] Guo, 2015.

[65] McCaslin and Erickson, 2019, p. 136.

vessels that the PLA Navy vessels are charged with protecting.[66] Somewhat surprisingly, a major challenge at least during the early stages was communicating with PRC-flagged civilian ships. According to PLA Navy Rear Admiral Xiao Xinnian, Chinese commercial vessels in the region desiring a naval escort were supposed to instead direct their requests via the PRC Ministry of Transportation back in Beijing.[67]

These Far Seas missions, while extremely modest by U.S. Navy standards, are providing invaluable training and "the most intense operational experience currently available to China's navy."[68] This reality is underscored by the fact that the PLA Navy staff has taken considerable effort to rotate through multiple ships, crews, and officers since China's first Gulf of Aden counter-piracy operations were dispatched in December 2008.[69] These rotations result in "the transfer of knowledge, skills, and perspective that occurs within China's navy upon the return of each escort group, increasing overall competency and professionalism."[70] PLA Navy officers, including prominent ship commanders and naval aviators, routinely serve multiple tours in the Gulf of Aden, and experience in the region is a significant factor in officer promotions.[71] How these rotations affect concepts and practice of control and command is a topic that merits further research.

In addition to the influence of the experience gained throughout the force in these deployments, these operations themselves have

---

66 Kristen Gunness and Samuel K. Berkowitz, "PLA Navy Planning for Out of Area Deployments," in Andrew Scobell, Arthur S. Ding, Phillip C. Saunders, and Scott W. Harold, eds., *The People's Liberation Army and Contingency Planning in China*, Washington, D.C.: National Defense University Press, 2015, p. 337.

67 Gunness and Berkowitz, 2015, p. 337.

68 Erickson and Strange, 2013, p. 2.

69 As of December 24, 2018, "the PLAN [PLA Navy] had sent 31 escort fleets, 100 ships, 67 shipboard helicopters, and more than 26,000 personnel to escort more than 6,600 PRC and foreign ships—in roughly equal proportion" to the Gulf of Aden. These deployments included 31 task forces as of January 2019. Andrew S. Erickson, "The China Anti-Piracy Bookshelf: Statistics and Implications from Ten Years' Deployment . . . and Counting," *Andrew S. Erickson: China Analysis from Original Sources*, blog post, January 2, 2019.

70 Erickson and Strange, 2013, p. 26.

71 Erickson and Strange, 2013, p. 27. We thank Nan Li for providing this information.

almost certainly had a major impact on how the PLA Navy assesses its chain of command and its control and command philosophy, particularly in terms of how micromanagement and communications issues might challenge their power projection efforts. For example, according to Wang Sheqiang, who later commanded the PLA Navy's contribution to Rim of the Pacific (RIMPAC) Exercise 2016,[72] operations in the Gulf of Aden saw a shift from "the fleet exercising command and the flotilla supporting to the flotilla exercising command and the fleet supporting," bringing advantages in flexibility and efficiency that have resulted in maritime control and command moving toward actual combat.[73] The translated *Global Times* article further explained,

> Upon entering the navigation phase, the reporter discovered changes in the "maritime command post." Before, the command post of the Far Seas training detachment could be generally described as "high in rank, large in scope, and commanded from above." "High-ranking" refers to command by flag officers, with senior colonels forming the main body [of the officers]. "Large in scope" refers to how there was a billet for every function, with tens of personnel. "Commanded from above" refers to how command authority of missions often rested with higher organizations despite involving mere detachment of vessels. Supposedly, this sort of jack-of-all-trades command post structure originated from the ground forces. Even though it carried many capabilities across many functions, the Far Seas deployment revealed problems such as complex command hierarchies and long preparation times. This time, the command post staff only consisted of several people, with an average age of 36. Most of them were lieutenant commander and captain–level officers. "This time we were only assigned a single meteorological forecasting personnel

---

72 RIMPAC is a large-scale multinational, multi-domain exercise intended to focus on maritime training, relationship-building, and interoperability among participant nations.

73 Guo Yuandan and Ju Zhen Hua, "While Under Close Reconnaissance by Foreign Forces, the Chinese Navy Far Seas Training Fell Was Wished a Happy Voyage [中国海军远海训练遭外军抵近侦察 被祝航行愉快]," *Global Times*, August 4, 2015.

> to strengthen support, from beginning to end, the detachment exercised independent command."[74]

Echoing Wang's comments, the retired PLA Navy rear admiral and military commentator Yin Zhuo opined that escort operations in the Gulf of Aden have driven the reform and innovation of PLA Navy command methods, with lessons learned already being applied to Far Seas training and other tasks.[75] To the extent that PLA Navy culture is evolving in ways that either constrain or enhance the PLA Navy's ability to project power far beyond China's shores, we expect that these missions can provide critical insight.

Example D: Far Seas Port

A landmark event in 2017 was the establishment of the PRC's first official overseas military base in the small African state of Djibouti, on the shores of the Red Sea. The establishment of this base required a substantial shift in PRC thinking. Beijing has long prided itself on being different from other great powers in that it did not station troops abroad. The official impetus behind the construction of the Djibouti base is as a logistics hub to support the Gulf of Aden task force operating nearby.[76] It is also useful to support PLA and People's Armed Police forces serving in United Nations peacekeeping operations in the Middle East and Africa.

However, the full significance of basing in Djibouti is that it marks a great leap outward in PRC and PLA thinking about global power projection.[77] It is evidence of a growing realization that China

---

[74] Guo and Ju, 2015.

[75] Feng Sheng and Yang Zurong, "Innovate and Strengthen, Then Set Sail Again [创新图强再起航]," *Liberation Army Daily*, February 26, 2016.

[76] Erica Downs, Jeffrey Becker, and Patrick deGategno, *China's Military Support Facility in Djibouti: The Economic and Security Dimensions of China's First Overseas Base*, Alexandria, Va.: CNA, 2017, p. 24.

[77] The term *great leap outward* has been used by one of the authors to characterize the dramatic results of China's post-Mao policy of reform and opening to the outside world and the accompanying substantial shift in PRC leadership outlook. See Andrew Scobell, "Introduction," in Andrew Scobell and Marylena Mantas, eds., *China's Great Leap Outward: The Hard*

cannot hope to improve its ability to project and sustain out-of-area military power without establishing an infrastructure or support network. According to Professor Liu Wanxia of the National Defense University's National Security Institute, "overseas bases not only provide reliable support for the overseas operations of major countries, but also greatly expand the coverage of military forces, strengthening response capabilities."[78] Beijing describes Djibouti as a "support base" and has focused on developing the 90-acre facility as a logistics hub. Nevertheless, considerable effort has been made to improve its capabilities to dock all but the two largest vessels in the PLA Navy's inventory. The base also has constructed a short runway and a helipad. It reportedly has a garrison of 400 uniformed personnel, including a detachment of PLA Marines.[79]

China's enduring ability to project and sustain naval power into the Far Seas will be dependent on the availability of logistics facilities and naval bases overseas. The very public establishment of the PLA's first official military base in Djibouti signifies the breaching of an important psychological barrier and likely heralds the creation of additional overseas bases in coming years. Subsequent locations for PLA bases are likely to be quite controversial: Probable candidates are in

---

and Soft Dimensions of a Rising Power, New York: Academy of Political Science, 2014, pp. 5–6. While this earlier usage referred to economic and diplomatic spheres, here the term is purposefully used to refer to the military domain and changes in Chinese military thinking. In other words, Beijing has belatedly recognized that China's burgeoning overseas interests require protection, and a significant part of the answer is to be found in improved global power projection by the PLA. See Andrew Scobell and Nathan Beauchamp-Mustafaga, "The Flag Lags but Follows: The PLA and China's Great Leap Outward," in Phillip C. Saunders, Arthur S. Ding, Andrew Scobell, Andrew N. D. Yang, and Joel Wuthnow, eds., *Chairman Xi Remakes the PLA*, Washington, D.C.: National Defense University Press, 2019. See also Jonas Parello-Plesner and Mathieu Duchâtel, *China's Strong Arm: Protecting Citizens and Assets Abroad*, London: International Institute for Strategic Studies, 2015.

[78] Liu Wanxia [刘万侠], "Historical Reflections on Major Countries Dispatching Military Forces Overseas [世界大国 '军事力量走出去' 的历史思考]," *Liberation Army Daily*, April 6, 2017. See also Xue Guifang [薛桂芳] and Zheng Jie [郑洁], "The Present Needs and Potential Threats of China's Overseas Base Construction in the 21st Century [中国21世纪海外基地建设的现实需求与风险应对]," *Global Review* [国际展望], No. 4, 2017, pp. 104–121.

[79] Scobell and Beauchamp-Mustafaga, 2019, p. 189.

South Asia, Southeast Asia, and the South Pacific.[80] Of course, China does not need to establish and operate its own network of bases; it merely needs ready access to the facilities of friendly states. Nevertheless, Djibouti is likely to become a desirable model for future overseas bases. Djibouti is also likely to play a significant role in the expanding role and power projection aspirations of the PLA's growing Marine Corps.

As additional Chinese bases overseas are developed, as is expected, it is unknown how China's centralized concept of control and command will continue be executed. The PLA Navy personnel who operate the base report directly to the PLA Navy staff, who in turn report to the CMC's Joint Staff Office. As with Far Seas operations, the chain of command in Far Base operations bypasses the theater commands.[81] Further complicating this chain of command is that the PLA Marines stationed at the base in Djibouti have their own separate headquarters element and distinct control and command structure.[82] This construct could create regional challenges and invite greater micromanagement, given the strategic implications of operations conducted so far from Chinese waters. However, becoming more distributed globally will almost certainly require greater delegation of certain decisionmaking responsibilities, given the challenges posed by distance, time zones, and communications architecture.

The Future of Control and Command in PLA Navy Power Projection

During recent decades, China has demonstrated the ability to project and sustain significant maritime power in the Near Seas—relying on

---

[80] Sumit Ganguly and Andrew Scobell, "The Himalayan Impasse: Sino-Indian Rivalry in the Wake of Doklam," *Washington Quarterly*, Vol. 41, No. 3, Fall 2018, p. 183.

[81] Saunders, 2019, p. 17.

[82] See, for example, Dennis J. Blasko and Roderick Lee, "The Chinese Navy's Marine Corps, Part 2: Chain-of-Command Reforms and Evolving Training," *China Brief* (Jamestown Foundation), Vol. 19, No. 4, February 15, 2019.

the control and command system. More recently, China has also displayed improved ability to project modest levels of naval power into the Far Seas. Small task forces have learned to operate effectively at considerable distances from home ports and sustain their presence for many months at a time. Consecutive counter-piracy task forces in the Gulf of Aden since December 2008 have provided invaluable experience in PLA Navy control and command. Although these have been significant experiences, they have also been limited noncombat missions with small flotillas. Exercising control and command over a larger number of vessels—such as in the carrier groups that have conducted drills with the *Liaoning* in recent years inside and just beyond the First Island Chain—offers a far greater challenge but potentially also much larger lessons.

The PLA Navy is keen to let other maritime agencies take the lead on "maritime rights enforcement" but cannot avoid the headaches of interagency coordination. To be sure, Beijing has decided that the Coast Guard and maritime militia will be on the front lines as the PRC deals with day-to-day sovereignty disputes against the blue hulls, white hulls, and gray hulls from other countries in the South China and East China Seas. In the Near Seas, the PLA Navy is content, as others have observed, to "lead from behind," displaying no great enthusiasm to lead from in front.[83] But PLA Navy leaders are keen to help train personnel from maritime law enforcement services so that these agencies can effectively execute their duties.[84] The lead agency in this effort is the Coast Guard, and white hulls have become China's "blunt defenders of sovereignty."[85]

The introduction of a sizable number of new platforms, advances in information technology, and high-tech military weaponry together provide the PLA Navy with unprecedented potential for enhanced operational capabilities. Moreover, the PLA's evolving doctrine theoretically permits the armed forces to demonstrate greater initiative and to execute more-complex operations. Nevertheless, operators continue

---

[83] McCaslin and Erickson, 2019, pp. 147–148.

[84] Jakobson, 2015, pp. 311 and 315.

[85] Morris, 2017.

to be restrained by superiors who seek to micromanage their subordinates and are inclined to make tactical-level decisions. This is the implicit criticism of Director Bao Daohua of the East China Sea Fleet Staff. Bao told reporters in July 2017: "The lower the level of command, the stronger our commanding ability is, and the more we can adapt to the needs of operations."[86] The PLA's traditional rigid chain of command structure and a constricting control and command culture tend to inhibit prompt decisionmaking and hamper swift actions—both critical components of the power projection mission.

Indeed, the heavy demands of a range of power projection missions in the Near and Far Seas have likely strained this structure and culture. The above examples suggest that the PLA's traditional chain of command structure and the culture of control and command could hamper mission success. What remains unclear is whether China's sailors are achieving mission success by devising ad hoc modes of operation and developing temporary command workarounds, or if instead the PLA Navy's culture over the past decade has been influenced by increased Near Seas and Far Seas operational experience. This latter explanation is compelling to us, but there is insufficient hard evidence to make a determination; nevertheless, it would help explain the largely impressive performance of the PLA Navy in the four peacetime power projection examples we surveyed.[87] The fact that Chinese naval officers and enlisted personnel have been able to execute their missions in spite of structural and cultural hindrances indicates to us that PLA Navy personnel have adapted command structures and evolved the control and command culture to survive and succeed.[88] However, as operations expand into more-contested environments, we expect that such adaptation likely would not endure beyond these more limited circumstances.

---

[86] Liu Yaxun [刘亚迅], Wang Junshuo [王凌硕], and Chen Guoquan [陈国全], "Far Sea Ocean Sharp Sword Unsheathed [远海大洋利剑出鞘]," *China Military Net*, July 24, 2017.

[87] Other analysts share this assessment. See, for example, McCaslin and Erickson, 2019, p. 153.

[88] For a recent comprehensive assessment of PLA Navy personnel recruitment, education, and training, see Kenneth Allen and Morgan Clemens, *The Recruitment, Education, and Training of PLA Navy Personnel*, Newport, R.I.: U.S. Naval War College, China Maritime Studies Institute, 2014.

Nevertheless, the significance of this accomplishment should neither be underestimated nor exaggerated. On one hand, these developments represent a substantial step forward for the PLA Navy and attest to its enhanced capabilities in projecting power at a range of distances from China's shores and improved ability to operate in complex conditions. On the other hand, this achievement has thus far been limited to naval operations in mostly permissive environments. Operating in wartime or during high-intensity conflicts would very likely be a fundamentally different story. A contingency with Taiwan, for example, could heavily stress control and command processes in the PLA Navy and across the PLA joint force, particularly if a third force—such as the U.S. Navy—were also participating.

In wartime, high-level civilian and military superiors in the PLA Navy's chain of command would likely be far more unwilling to overlook or accept violations of the chain of command or deviations from the control and command culture on the part of subordinates. Although there is the real risk of mission failure or degraded mission outcomes in any future fight, there is also a possibility of culture change that could shift how control and command is executed, albeit slowly. This could play out differently in different environments: For example, mission failure might be more conceivable in the Near Seas than in the Far Seas because the former has more layers of command and components involved, such as interagency actors and theater commands. However, to date the PLA has yet to stand up "a standing or ad hoc Joint Task Force mechanism" to command Far Seas operations.[89] Such a structure might further complicate control and command issues in a Far Seas contingency and add additional layers of centralized control.

It is an open question whether or not the PLA Navy will be able to leverage the experiences of coordinating with PRC maritime paramilitary formations in the Near Seas and working with state-owned enterprises, PRC commercial vessels, and other civilian bureaucracies, including the Ministry of Foreign Affairs, in the Far Seas. Would the procedures and protocols devised for the intricate cabbage strategy in the South China Sea, whereby PLA Navy ships coordinate with Coast

[89] Saunders, 2019, p. 16.

Guard vessels and maritime militia boats, work at greater distances from China? Would the protocols formulated for working with PRC civilian commercial oil tankers and container ships in and around the Gulf of Aden be readily applicable in other situations? Would procedures formulated by the PLA Navy for the overseas evacuation of PRC citizens in operations such as the one from Yemen in 2015 facilitate future cooperation with ostensibly civilian entities, assets, and platforms overseas?[90] Although the CCP has taken steps to formalize civil-military cooperation to support maritime capabilities,[91] the answers to these questions are difficult to predict.

The operations in the Near Seas where the PLA Navy coordinates very closely with the Coast Guard and maritime militia officially involves tight control and command structures. Unofficially, there is likely to be considerable informal flexibility because personnel from the other agencies are not as well trained as PLA sailors, and they lack the experience and level of discipline of military professionals. Absent flexibility by the PLA Navy, the coordination might not work. In Far Seas operations, we might expect the chain of command to become looser and more flexible over time because, despite tendencies toward micromanagement in the near term, commanders at sea become more comfortable with making operational command decisions on their own. The combined impact of these experiences could produce a PLA Navy that is increasingly adaptive to different conditions in power projection missions, including how they are commanded and controlled, which would represent a cultural shift with operational and strategic implications.

[90] Chase, 2015, pp. 312–313.

[91] Jeffrey Becker, Erica Downs, Ben DeThomas, and Patrick deGategno, *China's Presence in the Middle East and Western Indian Ocean: Beyond Belt and Road*, Alexandria, Va.: CNA, February 2019.

CHAPTER FOUR

Key Questions

In the preceding chapters, we identified how the starkly different concepts of C2 in the U.S. Navy and the PLA Navy might affect each service's ability to shift to counter–power projection and power projection missions, respectively. Overall, we found that both the U.S. Navy and the PLA Navy will likely be challenged to fully shift to these new strategic postures in various ways if they do not adapt their existing concepts of C2. For the U.S. Navy, its model of mission command appears conducive to counter–power projection missions in theory, but success will likely require increased investments in education and professionalism across the force. The PLA Navy's rigid control and command structure, which endures even as its maritime operations have evolved, is likely to come under increasing strain given the relative independence and greater operations tempo required by power projection operations.

Together, these challenges raise critical questions about if and how both navies will adapt their culturally ingrained C2 structures to accommodate these strategic shifts to new missions. Each navy's willingness to adapt could prove to be decisive in maritime competition, and perhaps ultimately in the balance of strategic competition between the United States and China overall. Currently, many unknowns exist, particularly in understanding how PLA Navy culture is evolving and how the CCP might weigh its preferred method of tight control throughout the PLA against more-effective power projection efforts.

Rather than seeking to identify definitive answers about these topics, this research was intended to uncover key areas of inquiry that we believe will provide the greatest insights into the balance of naval competition between the United States and China in the coming years. Specifically, we focused on identifying what questions about the PRC's shift toward power projection require additional examination. Although many analyses exist about the "hard power" dimensions of the PLA, comparatively few provide insight into the cultural and organizational aspects of PLA capabilities. As a result, there is much greater understanding of what China can *employ* in a wartime scenario, and relatively little is known about *how effectively* the PLA would be able to perform. Therefore, we identified the following four questions we hope will inform future analysis.

- **What is more valuable to the PRC: the ability to project power globally or retaining its rigid control and command system?** These two goals are in inherent tension with one another, which will likely increase as the PLA Navy expands its power projection missions further and begins to operate in more-contested environments. Although we do know that one will most likely have to give for the other to advance, we do not know which the PRC is more likely to prioritize. However, if it chooses to prioritize control and command, this does not necessarily mean that the PLA Navy would fail in its power projection mission. It could mean that the PLA Navy might choose to focus on operations inside the First Island Chain, which would lend itself to greater direct control.
- **Will the PLA Navy's increased experience and professional development affect the trust placed in PLA Navy personnel by senior PLA commanders? And how will increased PLA Navy professionalism affect control and command?** Through its current operations, the PLA Navy is gaining valuable experience that in theory could generate greater professionalism and trust from senior commanders to enable more of a mission command philosophy. However, even if greater training and experience occurs, that does not necessarily mean that the PLA culture will allow this to translate to greater trust in a subordinate's ability to operate and think independently or to less centralization. Further, gaining additional technical skills and wartime exposure develop a certain type

of experience, but that is not the same as developing and training personnel to practice and value empowered execution and articulation of commander's intent, for example.

- **Would the CCP tolerate a PLA Navy that is more empowered to make independent decisions?** A service that has greater autonomy and freedom of action is by definition more independent and powerful than one that is tightly and directly controlled. This is particularly salient as it relates to the PLA Navy's submarine forces, which are developing a fleet of nuclear-powered submarines outfitted with ballistic missiles. The top echelon of PRC civilian leaders and PLA leaders will likely exercise particularly tight control and command where nuclear weapons are concerned—especially regarding launch command, as senior CCP leaders in particular have always taken a special interest and direct involvement in China's nuclear weapons program.[1] This could result in different methods of control and command depending on the type of military capability. A related question is whether a distinct subculture could emerge within the PLA Navy, as it shows signs of becoming the most forward-leaning service in terms of relative operational independence (such as in its operation of an overseas base in Djibouti). Such a subculture could eventually challenge the PRC's concept of complete CCP control of the PLA or, alternatively, could serve the PRC's power projection purposes at the same time as becoming more distinct from the PLA.
- **Would the PLA Navy taking a mission command approach to C2 be a threat to the United States?** An underlying assumption we identified throughout the course of our research is that

[1] John Wilson Lewis and Xue Litai, *China Builds the Bomb*, Stanford, Calif: Stanford University Press, 1988. For more about the significant autonomy granted to the PLA on purely military matters by civilian political leaders, see Phillip C. Saunders and Andrew Scobell, "Introduction: PLA Influence on National Security Policymaking," in Phillip C. Saunders and Andrew Scobell, eds., *PLA Influence on China's National Security Policymaking*, Stanford, Calif.: Stanford University Press, 2015. On significant senior civilian leadership interest in nuclear matters, including doctrine, as distinct from conventional warfighting matters, see M. Taylor Fravel, *Active Defense: China's Military Strategy Since 1949*, Princeton, N.J.: Princeton University Press, 2019, pp. 236–269.

the PLA's power projection efforts are hindered by strict control and command and would therefore be bolstered by a transition to a mission command culture. Although this might be true, we do not have the evidence to support this assumption. Even if the PLA Navy were able to shift to a more decentralized C2 construct, it might not closely resemble what is practiced in the U.S. Navy. The PRC might choose to emphasize power projection while shifting to an entirely different C2 system that could pose an even greater threat.

Conclusion

Although the PLA Navy has demonstrated an impressive ability to conduct new missions despite its constricting control and command system, as well as leverage "asymmetric" assets and capabilities, operating in wartime would present added complexities and stresses. If future PLA Navy operations fall outside designated areas of responsibility of any of the five theater commands, they would likely present significant cultural and structural challenges to any mission. Whether the PRC's valuation of rigid C2, or its power projection missions, would assume more risk is the key question.

For the U.S. Navy, the shift to counter–power projection could increase the practice of mission command within the U.S. Navy, because of a more distributed posture and the need to further decentralize decisionmaking to support sea control missions. However, the U.S. Navy will likely have to address structural barriers, such as leader development, legacy force structure and operating concepts, and its orientation toward traditional warfare over asymmetric competition. The extent to which the U.S. Navy addresses these barriers will determine the degree of impact that the shift to counter–power projection has on the U.S. Navy's command culture and its ability to exercise mission command.

Although many unknowns exist regarding the future balance of maritime competition between the United States and China, it is clear that both nations will be challenged to conform, and possibly adapt their cultures, to the requirements of the new strategic orientation. Although the differing concepts of C2 in both navies offer a window into how these shifts might occur, several variables that require further observation will stand to affect the competitive playing field. The themes and related questions posed in this report can help us to understand these changes as they evolve.

References

Allen, Kenneth, "Introduction to the People's Liberation Army's Administrative and Operations Structure," in James C. Mulvenon and Andrew N. D. Yang, eds., *The People's Liberation Army as Organization: Reference Volume v1.0*, Santa Monica, Calif.: RAND Corporation, CF-182-NSRD, 2002, pp. 1–43. As of June 8, 2020: https://www.rand.org/pubs/conf_proceedings/CF182.html

Allen, Kenneth, and Morgan Clemens, *The Recruitment, Education, and Training of PLA Navy Personnel*, Newport, R.I.: U.S. Naval War College, China Maritime Studies Institute, 2014.

Bayer, Michael, and Gary Roughead, *Strategic Readiness Review*, Washington, D.C.: U.S. Navy, December 3, 2017, pp. 37–62.

Becker, Jeffrey, Erica Downs, Ben DeThomas, and Patrick deGategno, *China's Presence in the Middle East and Western Indian Ocean: Beyond Belt and Road*, Alexandria, Va., CNA, February 2019. As of March 23, 2020: https://www.cna.org/CNA_files/PDF/DRM-2018-U-018309-Final2.pdf

Berger, David, *Commandant's Planning Guidance*, Washington, D.C., U.S. Marine Corps: Washington, D.C., 2019.

Blasko, Dennis J., *The Chinese Army Today: Tradition and Transformation in the 21st Century*, 2nd ed., New York: Routledge, 2012.

———, "The Chinese Military Speaks to Itself, Revealing Doubts," *War on the Rocks*, February 18, 2019. As of February 17, 2020: https://warontherocks.com/2019/02/the-chinese-military-speaks-to-itself-revealing-doubts/

Blasko, Dennis J., and Roderick Lee, "The Chinese Navy's Marine Corps, Part 2: Chain-of-Command Reforms and Evolving Training," *China Brief* (Jamestown Foundation), Vol. 19, No. 4, February 15, 2019. As of March 23, 2020: https://jamestown.org/program/the-chinese-navys-marine-corps-part-2-chain-of-command-reforms-and-evolving-training/

Boyd, John, "Organic Design for Command and Control," in Grant T. Hammond, ed., *A Discourse on Winning and Losing*, Maxwell, Ala.: Air University Press, 2018, pp. 218–240.

Brender, L. Burton, "The Problem of Mission Command," *Strategy Bridge*, September 1, 2016. As of January 29, 2020:
https://thestrategybridge.org/the-bridge/2016/9/1/the-problem-of-mission-command

Builder, Carl H., *The Masks of War: American Military Styles in Strategy and Analysis*, Baltimore, Md.: Johns Hopkins University Press, 1989.

"Carrier Leads Exercise in West Pacific," *China Daily*, April 23, 2018. As of March 30, 2020:
http://eng.chinamil.com.cn/view/2018-04/23/content_8011325.htm

Chan, Minnie, "Chinese Navy Trains Top Guns to Command Expanding Aircraft Carrier Fleet," *South China Morning Post*, December 11, 2019a. As of December 17, 2019:
https://www.scmp.com/news/china/military/article/3041695/chinese-navy-trains-top-guns-command-expanding-aircraft-carrier

———, "First Made-in-China Aircraft Carrier, the Shandong, Officially Enters Service," *South China Morning Post*, December 17, 2019b.

Chappell, Bill, "China Reaches Deal to Build Military Outpost in Djibouti," *Two-Way* (NPR), January 21, 2016. As of March 28, 2020:
http://www.npr.org/sections/thetwo-way/2016/01/21/463829799/china-reaches-deal-to-build-military-outpost-in-djibouti

Chase, Michael, "The PLA and Far Sea Contingencies: Chinese Capabilities for Noncombatant Evacuation Operations," in Andrew Scobell, Arthur S. Ding, Phillip C. Saunders, and Scott W. Harold, eds., *The People's Liberation Army and Contingency Planning in China*, Washington, D.C.: National Defense University Press, 2015, pp. 301–319.

Chase, Michael S., Jeffrey Engstrom, Tai Ming Cheung, Kristen Gunness, Scott Warren Harold, Susan Puska, and Samuel K. Berkowitz, *China's Incomplete Military Transformation: Assessing the Weaknesses of the People's Liberation Army (PLA)*, Santa Monica, Calif.: RAND Corporation, RR-893-USCC, 2015. As of February 2, 2020:
https://www.rand.org/pubs/research_reports/RR893.html

Clark, Bryan, Adam Lemon, Peter Haynes, Kyle Libby, and Gillian Evans, *Regaining the High Ground at Sea: Transforming the U.S. Navy's Carrier Air Wing for Great Power Competition*, Washington, D.C.: Center for Strategic and Budgetary Assessments, 2018.

Clark, Bryan, and Jesse Sloman, *Advancing Beyond the Beach: Amphibious Operations in an Era of Precision Weapons*, Washington, D.C.: Center for Strategic and Budgetary Assessments, 2016.

Cliff, Roger, *China's Military Power: Assessing Current and Future Capabilities*, New York: Cambridge University Press, 2015.

Cole, Bernard D., *The Great Wall at Sea: China's Navy in the Twenty-First Century*, 2nd ed., Annapolis, Md.: Naval Institute Press, 2010.

Crane, Conrad, "Mission Command and Multi-Domain Battle Don't Mix," *War on the Rocks*, August 23, 2017. As of February 3, 2020: https://warontherocks.com/2017/08/mission-command-and-multi-domain-battle-dont-mix/

Downs, Erica, Jeffrey Becker, and Patrick deGategno, *China's Military Support Facility in Djibouti: The Economic and Security Dimensions of China's First Overseas Base*, Alexandria, Va.: CNA, 2017.

Englehorn, Lyla, *Distributed Maritime Operations (DMO) Warfare Innovation Continuum (WIC) Workshop September 2017: After Action Report*, Monterey, Calif.: Naval Postgraduate School, Consortium for Robotics and Unmanned Systems Education and Research, November 2017.

Episkopos, Mark, "Carrier Superpower? How Many Aircraft Carriers Does China Really Want?" *National Interest*, August 15, 2019. As of January 30, 2020: https://nationalinterest.org/blog/buzz/carrier-superpower-how-many-aircraft-carriers-does-china-really-want-73591

Erickson, Andrew S., "The China Anti-Piracy Bookshelf: Statistics and Implications from Ten Years' Deployment . . . and Counting," *Andrew S. Erickson: China Analysis from Original Sources*, January 2, 2019. As of March 23, 2020: http://www.andrewerickson.com/2019/01/the-china-anti-piracy-bookshelf-statistics-implications-from-ten-years-deployment-counting/

Erickson, Andrew S., and Conor M. Kennedy, "China's Maritime Militia: What It Is and How to Deal with It," *Foreign Affairs*, June 23, 2016. As of January 30, 2020: https://www.foreignaffairs.com/articles/china/2016-06-23/chinas-maritime-militia

Erickson, Andrew S., and Austin M. Strange, *No Substitute for Experience: Chinese Anti-Piracy Operations in the Gulf of Aden*, Newport, R.I.: Naval War College China Maritime Studies Institute, November 2013.

———, *Six Years at Sea . . . and Counting: Gulf of Aden Anti-Piracy and China's Maritime Commons Presence*, Washington, D.C.: Jamestown Foundation, June 2015.

Fan Yihua [范印化], "Vigorously Promote the Greater Development and Greater Glory of Naval Culture [大力推进海军文化大发展大荣誉]," *Qiushi*, June 30, 2009. As of December 18, 2019: http://www.qstheory.cn/zxdk/2008/200811/200906/t20090609_1287.htm

Feng Sheng and Yang Zurong, "Innovate and Strengthen, Then Set Sail Again [创新图强再起航]," *Liberation Army Daily*, February 26, 2016. As of March 28, 2020: http://www.81.cn/jfjbmap/content/2016-02/26/content_2493.htm

Fravel, M. Taylor, *Active Defense: China's Military Strategy Since 1949*, Princeton, N.J.: Princeton University Press, 2019.

Ganguly, Sumit, and Andrew Scobell, "The Himalayan Impasse: Sino-Indian Rivalry in the Wake of Doklam," *Washington Quarterly*, Vol. 41, No. 3, Fall 2018.

Gilday, Michael, *FRAGO 01/2019: A Design for Maintaining Maritime Superiority*, Washington, D.C.: U.S. Navy, December 2019.

Gunness, Kristen, and Samuel K. Berkowitz, "PLA Navy Planning for Out of Area Deployments," in Andrew Scobell, Arthur S. Ding, Phillip C. Saunders, and Scott W. Harold, eds., *The People's Liberation Army and Contingency Planning in China*, Washington, D.C.: National Defense University Press, 2015, pp. 321–347.

Guo Yuandan [郭媛丹], "Foreign Forces Engage in Close Reconnaissance of Chinese Ships During Far Seas Training, Express Pleasant Journey [中国海军远海训练遭外军抵近侦察被祝航行愉快]," *Global Times*, August 4, 2015. As of December 18, 2019:
http://www.xinhuanet.com//mil/2015-08/04/c_128090403_3.htm

Guo Yuandan and Ju Zhen Hua, "While Under Close Reconnaissance by Foreign Forces, the Chinese Navy Far Seas Training Fell Was Wished a Happy Voyage [中国海军远海训练遭外军抵近侦察 被祝航行愉快]," Global Times, August 4, 2015. As of March 28, 2020:
https://mil.huanqiu.com/article/9CaKrnJO8bp

Hill, Andrew, and Heath Niemi, "The Trouble with Mission Command: Flexive Command and the Future of Command and Control," *Joint Force Quarterly*, Vol. 86, No. 3, June 21, 2017.

Jakobson, Linda, "The PLA and Maritime Security Actors," in Phillip C. Saunders and Andrew Scobell, eds., *PLA Influence on China's National Security Policymaking*, Stanford, Calif.: Stanford University Press, 2015, pp. 300–323.

Joint Military Operations Department, *Syllabus and Study Guide for the Joint Maritime Operations Intermediate Lever Warfighter's Course*, Newport, R.I.: U.S. Naval War College, College of Naval Command and Staff and Naval Staff College, February 2018.

Joint Publication 3-32, *Command and Control of Joint Maritime Operations*, Washington, D.C.: Joint Chiefs of Staff, August 7, 2013.

Joint Publication 3-32, *Command and Control of Joint Maritime Operations*, Washington, D.C.: Joint Chiefs of Staff, June 8, 2018.

Kaufman, Alison A., and Peter W. Mackenzie, *Field Guide: The Culture of the Chinese People's Liberation Army*, Alexandria, Va.: CNA, 2009.

King, Ernest, Commander in Chief, Atlantic Fleet, U.S. Navy, "Exercise of Command: Excess of Detail in Orders and Instructions," CINCLANT Serial 053, January 21, 1941.

Larter, David B., "The U.S. Surface Fleet Is Shaking Up How It Readies Ships for Deployment," *Defense News,* May 6, 2019.

Larter David B., and Mark D. Faram, "Mattis Eyes Major Overhaul of Navy Deployments," *Navy Times,* May 7, 2018.

Ledberg, Sofia, "Rethinking the Political Control and Autonomy of the PLA," paper presented at the CAPS-RAND-NDU PLA Conference, Taipei, Taiwan, November 15–16, 2019.

Lewis, John Wilson, and Xue Litai, *China Builds the Bomb*, Stanford, Calif: Stanford University Press, 1988.

———, *Imagined Enemies: China Prepares for Uncertain War*, Stanford, Calif.: Stanford University Press, 2006.

Li Yimin and Li Xuefeng, "South Sea Fleet Aviation Division Relies on Information System to Boost the Level of Three Warfares Training," *Renmin Haijun*, September 2, 2014.

Lieberthal, Kenneth, *Governing China: From Revolution Through Reform*, 2nd ed., New York: W. W. Norton, 2004.

Liu Wanxia [刘万侠], "Historical Reflections on Major Countries Dispatching Military Forces Overseas [世界大国 '军事力量走出去' 的历史思考]," *Liberation Army Daily*, April 6, 2017. As of December 18, 2019:
http://www.xinhuanet.com//mil/2017-04/06/c_129525829.htm

Liu Yaxun [刘亚迅], Wang Junshuo [王凌硕], and Chen Guoquan [陈国全], "Far Sea Ocean Sharp Sword Unsheathed [远海大洋利剑出鞘]," *China Military Net*, July 24, 2017. As of December 18, 2019:
http://www.81.cn/jfjbmap/content/2017-07/24/content_183062.htm

McCaslin, Ian Burns, and Andrew S. Erickson, "The Impact of Xi-Era Reforms on the Chinese Navy," in Phillip C. Saunders, Arthur S. Ding, Andrew Scobell, Andrew N. D. Yang, and Joel Wuthnow, eds., *Chairman Xi Remakes the PLA*, Washington, D.C.: National Defense University Press, 2019, pp. 125–170.

McCullough, Amy, "Goldfein's Multi-Domain Vision," *Air Force Magazine*, August 29, 2018. As of January 29, 2020:
https://www.airforcemag.com/article/goldfeins-multi-domain-vision/

McDaniel, Darwin, "Navy Looks to Improve Training with Live, Virtual, Constructive Technologies," *Executive Gov*, November 28, 2018. As of January 30, 2020:
https://www.executivegov.com/2018/11/
navy-looks-to-improve-training-with-live-virtual-constructive-technologies/

Moore, Dale, and Gregory Smith, "The Navy Needs a Culture of Innovation," *Proceedings*, August 2019. As of February 3, 2020:
https://www.usni.org/magazines/proceedings/2019/august/
navy-needs-culture-innovation

Morris, Lyle J., "Blunt Defenders of Sovereignty: The Rise of Coast Guards in East and Southeast Asia," *Naval War College Review*, Vol. 70, No. 2, Spring 2017, pp. 75–112.

Naval Surface Force, U.S. Pacific Fleet, *Surface Force Strategy: Return to Sea Control*, Naval Amphibious Base Coronado, Calif., 2017. As of February 13, 2020: https://www.public.navy.mil/surfor/Documents/Surface_Forces_Strategy.pdf

O'Rourke, Ronald, "PLAN Force Structure: Submarines, Ships, and Aircraft," in Phillip C. Saunders, Christopher Yung, Michael Swaine, and Andrew Nien-Dzu Yang, eds., *The Chinese Navy: Expanding Capabilities, Evolving Roles*, Washington, D.C.: National Defense University Press, 2011.

———, *China's Naval Modernization: Implications for U.S. Navy Capabilities—Background and Issues for Congress*, Washington, D.C.: Congressional Research Service, December 20, 2019.

Panda, Ankit, "Power Plays Across the First Island Chain: China's Lone Carrier Group Has a Busy December," *The Diplomat*, December 27, 2016.

Parello-Plesner, Jonas, and Mathieu Duchâtel, *China's Strong Arm: Protecting Citizens and Assets Abroad*, London: International Institute for Strategic Studies, 2015.

Peck, Michael, "China Is Tripling the Size of Its Marine Corps," *National Interest*, November 2, 2019. As of March 31, 2020: https://nationalinterest.org/blog/buzz/china-tripling-size-its-marine-corps-92891

Perlmutter, Amos, and William M. LeoGrande, "The Party in Uniform: Toward a Theory of Civil-Military Relations in Communist Political Systems," *American Political Science Review*, Vol. 76, No. 4, December 1982, pp. 778–789.

Public Law 99-433, Goldwater-Nichols Department of Defense Reorganization Act of 1986, October 1, 1986.

Richardson, John, "The Navy Our Nation Needs," speech delivered at the Heritage Foundation, Washington, D.C., February 1, 2018a.

———, *The Charge of Command*, Washington, D.C.: U.S. Navy, April 6, 2018b.

———, *Navy Leader Development Framework*, Version 3.0, Washington, D.C.: U.S. Navy, May 2019.

Roberts, Colin, "The Navy," in S. Rebecca Zimmerman, Kimberly Jackson, Natasha Lander, Colin Roberts, Dan Madden, and Rebeca Orrie, *Movement and Maneuver: Culture and the Competition for Influence Among the U.S. Military Services*, Santa Monica, Calif.: RAND Corporation, RR-2270-OSD, 2019, pp. 47–75. As of June 8, 2020: https://www.rand.org/pubs/research_reports/RR2270.html

Saunders, Phillip C., "Beyond Borders: PLA Command and Control of Overseas Operations," paper presented at the CAPS-RAND-NDU PLA Conference, Taipei, Taiwan, November 15–16, 2019.

Saunders, Phillip C., Arthur S. Ding, Andrew Scobell, Andrew N.D. Yang, and Joel Wuthnow, eds., *Chairman Xi Remakes the PLA*, Washington, D.C.: National Defense University Press, 2019.

Saunders, Phillip C., and Andrew Scobell, "Introduction: PLA Influence on National Security Policymaking," in Phillip C. Saunders and Andrew Scobell, eds., *PLA Influence on China's National Security Policymaking*, Stanford, Calif.: Stanford University Press, 2015, pp. 1–32.

Saunders, Phillip C., and Joel Wuthnow, "China's Goldwater-Nichols? Assessing PLA Organizational Reforms," *Joint Force Quarterly*, Vol. 82, No. 3, July 2016, pp. 68–75.

Scarbro, Graham, "'Go Straight at 'Em!': Training and Operating with Mission Command," *Proceedings*, Vol. 145, No. 5, May 2019.

Scobell, Andrew, *China's Use of Military Force: Beyond the Great Wall and the Long March*, New York: Cambridge University Press, 2003.

———, "Discourse in 3-D: The PLA's Evolving Doctrine, Circa 2009," in Roy Kamphausen, David Lai, and Andrew Scobell, eds., *The PLA at Home and Abroad: Assessing the Operational Capabilities of China's Military,* Carlisle, Pa.: U.S. Army War College, 2010.

———, "Introduction," in Andrew Scobell and Marylena Mantas, eds., *China's Great Leap Outward: Hard and Soft Dimensions of a Rising Power,* New York: Academy of Political Science, 2014.

———, *Civil-Military "Rules of the Game" on the Eve of China's 19th Party Congress*, Seattle, Wash.: National Bureau of Asia Research, October 2017.

———, "The South China Sea and U.S.-China Rivalry," *Political Science Quarterly*, Vol. 133, No. 2, 2018.

Scobell, Andrew, and Nathan Beauchamp-Mustafaga, "The Flag Lags but Follows: The PLA and China's Great Leap Outward," in Phillip C. Saunders, Arthur S. Ding, Andrew Scobell, Andrew N. D. Yang, and Joel Wuthnow, eds., *Chairman Xi Remakes the PLA*, Washington, D.C.: National Defense University Press, 2019.

Scobell, Andrew, and Marylena Mantas, eds., *China's Great Leap Outward: Hard and Soft Dimensions of a Rising Power*, New York: Academy of Political Science, 2014.

Scobell, Andrew, Michael McMahon, and Cortez A. Cooper III, "China's Aircraft Carrier Program: Drivers, Developments, Implications," *Naval War College Review*, Vol. 69, No. 4, Autumn 2015, pp. 65–79.

Scobell, Andrew, and John M. Sanford, *North Korea's Military Threat: Pyongyang's Conventional Forces, Weapons of Mass Destruction, and Ballistic Missiles*, Carlisle, Pa.: U.S. Army War College, Strategic Studies Institute, 2007.

South, Todd, "The Army's Updated Warfighting Concept Will Drive Its Formations, Planning, and Experimentation," *Army Times*, December 6, 2018. As of January 30, 2020:
https://www.armytimes.com/news/your-army/2018/12/06/the-armys-updated-warfighting-concept-will-drive-its-formations-planning-and-experimentation/

State Council Information Office of the People's Republic of China, *China's National Defense in the New Era*, Beijing: Foreign Languages Press, July 2019.

Tangredi, Sam, "Antiaccess Warfare as Strategy," *Naval War College Review*, Vol. 71, No. 1, December 2018.

Thayer, Carl, "China's Oil Rig Gambit: South China Sea Game-Changer?" *The Diplomat*, May 12, 2014.

U.S. Army Training and Doctrine Command, *Multi-Domain Battle: Evolution of Combined Arms for the 21st Century: 2025–2040*, Fort Eustis, Va., December 2017.

U.S. Defense Intelligence Agency, *China Military Power: Modernizing a Force to Fight and Win*, Washington, D.C., 2019.

U.S. Navy Chief of Naval Operations, *A Design for Maintaining Maritime Superiority,* Version 2.0, Washington, D.C., December 2018. As of February 13, 2020:
https://www.navy.mil/navydata/people/cno/Richardson/Resource/Design_2.0.pdf

U.S. Office of Naval Intelligence, *The PLA Navy: New Capabilities and Missions for the 21st Century*, Suitland, Md., 2015.

Vandergriff, Donald E., *Adopting Mission Command: Developing Leaders for a Superior Command Culture*, Annapolis, Md.: U.S. Naval Institute Press, 2019.

Vangjel, Peter, "Mission Command: A Practitioner's Guide," in Donald Vandergriff and Stephen Webber, eds., *Mission Command: The Who, What, When, Where, and Why: An Anthology*, Vol. 2, CreateSpace Independent Publishing Platform, 2019, pp. 162–182.

Werner, Ben, "Navy Proposing Bold Changes in How It Teaches Sailors in New Education Plan," *USNI News*, February 12, 2019. As of February 3, 2020:
https://news.usni.org/2019/02/12/41094

West, Bing, "American Naval Initiative: the Next Time Around," Hoover Institution, November 20, 2019. As of January 30, 2020:
https://www.hoover.org/research/american-naval-initiative-next-time-around

Wortzel, Larry M., "The General Political Department and the Evolution of the Political Commissar System," in James C. Mulvenon and Andrew N. D. Yang, eds., *The People's Liberation Army as Organization: Reference Volume v1.0*, Santa Monica, Calif.: RAND Corporation, CF-182-NSRD, 2002, pp. 225–245. As of June 12, 2020:
https://www.rand.org/pubs/conf_proceedings/CF182.html

Wuthnow, Joel, *China's Other Army: The People's Armed Police in an Era of Reform*, Washington, D.C:, Center for the Study of Chinese Military Affairs, Institute for National Strategic Studies, National Defense University, April 2019.

Wuthnow, Joel, and Phillip C. Saunders, *Chinese Military Reforms in the Age of Xi Jinping: Drivers, Challenges, and Implications*, Washington, D.C.: Center for the Study of Chinese Military Affairs, Institute for National Strategic Studies, National Defense University, March 2017.

———, "Introduction: Chairman Xi Remakes the PLA," in Phillip C. Saunders, Arthur S. Ding, Andrew Scobell, Andrew N. D. Yang, and Joel Wuthnow, eds., *Chairman Xi Remakes the PLA*, Washington, D.C.: National Defense University Press, 2019, pp. 1–24.

Xie Zhijun [谢志军] and Li Wei [李伟], "An Initial Examination of Establishing Culture with Marine Corps Special Characteristics [海军陆战队特色文化建设初探]," *Political Work Journal* [政工学刊], No. 11, 2017.

Xue Guifang [薛桂芳] and Zheng Jie [郑洁], "The Present Needs and Potential Threats of China's Overseas Base Construction in the 21st Century [中国21世纪海外基地建设的现实需求与风险应对]," *Global Review* [国际展望], No. 4, 2017, pp. 104–121. As of December 18, 2019:
http://www.siis.org.cn/UploadFiles/file/20170829/201704008%20%20%E8%96%9B%E6%A1%82%E8%8A%B3.pdf

Zimmerman, S. Rebecca, Kimberly Jackson, Natasha Lander, Colin Roberts, Dan Madden, and Rebeca Orrie, *Movement and Maneuver: Culture and Competition for Influence Among the U.S. Military Services*, Santa Monica, Calif.: RAND Corporation, RR-2270-OSD, 2019. As of June 8, 2020:
https://www.rand.org/pubs/research_reports/RR2270.html